ESSENTIAL MANNERS
FOR COUPLES

Also from The Emily Post Institute

◆

Emily Post's Etiquette, 17th edition

Emily Post's Wedding Etiquette, fourth edition

Emily Post's Wedding Planner, third edition

Emily Post's Entertaining

Emily Post's The Etiquette Advantage in Business, second edition

Essential Manners for Men

Emily Post's The Gift of Good Manners

Emily Post's The Guide to Good Manners for Kids

Emily Post's Favorite Party & Dining Tips

ESSENTIAL MANNERS
FOR COUPLES

FROM SNORING AND SEX
TO FINANCES AND FIGHTING FAIR—
WHAT WORKS, WHAT DOESN'T, AND WHY

PETER POST

Collins
An Imprint of HarperCollinsPublishers

FIRST EDITION

Designed by Joel Avirom and Jason Snyder

Printed on acid-free paper.

Library of Congress Cataloging-in-Publication Data

Post, Peter.
Essential manners for couples : from snoring and sex to finances and fighting fair—what works, what doesn't, and why / Peter Post.—1st ed.
p. cm.
1. Couples. 2. Marriage. 3. Man-woman relationships. 4. Etiquette. I. Title.

HQ734.P783 2005
646.7e8—dc22   2005046088
ISBN-10: 0-06-077665-X

ISBN-13: 978-0-06-077665-7

05 06 07 08 09 / 10 9 8 7 6 5 4 3 2 1

Some people may wonder, isn't it a little weird including a special thank-you page to the same person in two books?

I don't think so.

Frankly, when I wrote *Essential Manners for Men*, I had no idea there would be another book, much less one on couples. The fact is, these are two areas where my wife has tremendous influence on me: as a man and as a partner. If you read *Essential Manners for Men*, there seems to be no question I needed some help on that subject. And as for this book, well, I couldn't possibly write about couples without constantly being influenced by what has gone on between my wife and me for 32 years.

My wife is really a coauthor of this book. Without her, without the experiences we have shared together, without the ups and downs we have traveled together, I couldn't have written it.

Throughout this book, she is known as "my wife." So here again, I take great pleasure in thanking my wife, Tricia.

ACKNOWLEDGMENTS

WHETHER AT WORK OR WITH FRIENDS OR MEETING NEW PEOPLE, I've repeatedly been surprised by the willingness of people to talk about what it's like to be a part of a couple. They shared with me their triumphs and their faux pas, and explained what made their relationships work and what caused trouble. To all of you, my thanks for your insights and your open honesty.

In addition, I want to thank a number of people whose stories are sprinkled throughout the book: Michelle Lambert, Dana Apgar, Nicole and Michael Atherton, Leigh and Peter Phillips, Virginia and Mac Keyser, Dana and Ray Murphy, Maureen and Bill Post, Libby and Bill Post, Patsy Kolbe, Angela Zagursky, Andy Champine, Toni Sciarra, and Katherine O'Moore Klopf, Katherine and Scott Cowles.

Special thanks go to Royce Flippin, who has provided help and counsel on every page of the book, not only through his editing but also by his willingness to question each piece of my advice.

I can't imagine bringing a book like this to fruition without the help, passionate enthusiasm, and editorial advice of Toni Sciarra, my editor at HarperCollins.

I want to thank the following for their help and encouragement in making this book possible: Greg Chaput, Elizabeth Upham Howell, Cindy Post Senning, Peggy Post, Anna Post, Lizzie Post, Katherine Cowles, and Dick Dorr.

Finally, I am deeply indebted to all the people who responded to the couples survey. It was so very obvious from their responses that they enjoy the blessings of successful relationships. My heartfelt thanks to you all for your candor and your willingness to share your experiences.

CONTENTS

Preface

TWO PEOPLE ATTRACTED TO EACH OTHER, WILLING TO SHARE time and space with each other, and committed to being with each other and to being there for each other—that's what a couple is. Age, sex, and living situation are immaterial. What *is* material and universal are the issues that these two people face as they interact with each other and as they interact, as a couple, with other people.

Now, here's another universal truth: Because etiquette is about building relationships, etiquette will always play a significant role in how successfully these two people interact, whether they realize it or not.

That's what this book is about. My hope is that readers will find one or two "ah-ha" moments in each chapter and lots of other advice that confirms how well (or not) they are doing at making choices that affect their coupledom.

As you read this book, revel in the things you do well, think about the things that make you wonder what you could be doing differently, and talk with your partner about everything.

If you do, you'll find the end result is that you and your partner are that much more confident and comfortable in your relationship and that much more in a position to build the best, longest-lasting relationship possible.

INTRODUCTION

That's When the Lightbulb Went On

RECENTLY, I WAS TALKING TO A WOMAN MUCH YOUNGER THAN I who was in a committed relationship. It was a Christmas party, and I was enjoying my conversation with her. How could I possibly not? A good-looking, fun, twenty-something female paying attention to me for more than thirty seconds? I was in heaven—and my wife knew it. I couldn't help noticing, though, that the twenty-something was rather agitated and that the focus of her agitation was her gentleman friend, who was talking to another very attractive younger woman.

When I innocently asked if there was a problem, with daggers in her eyes she replied, "It's been more than five minutes, and he's still talking to her!"

"So?"

Her answer surprised me. "That's too long," she said. She was clearly unhappy.

"But you've been standing here talking with me for at least ten minutes," I countered. Apparently, I wasn't in the same league as the attractive twenty-something talking to her gentleman friend.

The moral of the story is this: What appears to be innocent behavior to one person in a relationship can be seen as egregious behavior by the other partner. Each couple has to make their own boundaries, and each partner needs to respect those boundaries.

A little later, I noticed that the couple was standing together. She looked happy. He smiled at me. "Eight and one half minutes," he said to me under his breath.

"All you get is five," I replied. They still need to talk.

A few weeks later, I was browsing in the self-help section of a bookstore in my hometown of Burlington, Vermont, when I was struck by a common theme running through the books. Virtually all of the advice books focused on how to recover after having screwed up your life—especially your life with your significant other.

If all the people who are buying these books in droves had tried using etiquette to help solve their relationship problems, I thought to myself, *then maybe their relationships wouldn't need saving, and they wouldn't need these books.*

That's when the lightbulb went on.

I knew then and there that I wanted to write a book about etiquette for couples. After all, the whole point of etiquette is to help you build better relationships by learning how to avoid or defuse situations that could become problematic. And your relationship with your SO is the most important relationship in your life. Therefore, I realized, a book about etiquette for couples makes perfect sense: Instead of fixing problems *after* they happen, couples can use etiquette to prevent problems from cropping up in the first place.

I can hear the protests already: "Why worry about *etiquette* in a relationship? We just want to relax and enjoy ourselves!" But that's exactly what etiquette lets you do: relax and enjoy yourselves. Etiquette is not about being excessively or insincerely polite with each other, or living by a set of artificial rules. It's about enhancing and enriching your relationship by

✦ Ensuring that you always treat each other with consideration, respect, and honesty

✦ Encouraging you to think before you act, rather than acting without thinking and regretting it later

✦ Improving the way you communicate as a couple

✦ Helping you to consistently choose the thoughtful word
or action that resolves a situation by building up your
relationship, rather than tearing it down

A strong relationship doesn't just happen. It takes root and
grows because each partner consciously works at making the
other partner's life as pleasant, as enjoyable, and as fulfilling as
possible. The success of your relationship will be directly pro-
portional to how considerate, respectful, and honest each of you
is with the other—or, to put it another way, how much etiquette
you bring to your relationship.

When we become part of a couple, we trade independence for
the security of having a 24/7 companion. That's great—no more
having to play the dating game, no more wondering if the other
person is interested, no more worrying that you're screwing up
your chances with him or her if you call or don't call. The other
person is simply *there*, a part of you and your life, seeing you at
your best and at your worst, 24/7, no getting away from it. . . .

✦ When you roll over in bed, there she is, taking your
covers.

✦ When you walk into your home at the end of a long
workday, there he is, making a mess on the couch with
chips and dip.

✦ When you get up in the morning, there she is, on the
toilet or hogging the sink.

✦ When you go for your morning paper and cup of
coffee, there he is, reading the front page.

✦ When you rush home to catch the game on TV, there she
is, watching a chick flick and commanding the remote.

The Post Couples Survey

Much of the statistical information cited in this book comes from the Post Couples Survey—an online survey that was conducted over a two-month period in the fall of 2004. The Emily Post Institute received over 600 responses to the survey, in which couples shared their detailed thoughts on a wide range of issues. In addition, an independent research firm conducted a focus group for the Emily Post Institute on relationship issues and also did a series of qualitative individual interviews with people in committed relationships from different regions of the country and various age groups. The comments and opinions of the people interviewed appear throughout the book.

It takes steady work to blend the lives of two individuals into the life of one couple. Some of this work is done in the privacy of your home; other work takes place when the two of you are out in public. For this reason, *Essential Manners for Couples* is divided into two major areas: your private life as a couple, and your public life as a couple.

Private Life

The first part of the book, "The Two of You Together," looks at critical issues that affect the two of you in private, including how to communicate effectively, and what simple actions you can take to enhance your relationship. It draws a bead on why special days *are* special, and how you can improve your relationship by changing your approach to those red-letter dates. Your bedroom is perhaps the most vulnerable and intimate place you share in your lives together; as you'll see, a little etiquette can go a long way here. The book also examines how etiquette can improve other areas of your private life that affect your relationship significantly: children, finances, chores, disputes, even how you spend your leisure time.

What goes on behind closed doors defines who you are as a couple and how successful your relationship will be. As this section explains, if you make a practice of treating each other with consideration and respect, you can avoid a lot of potential conflicts. Even better, when there is a conflict—and there *will* be conflicts—this section offers strategies for resolving your issues before they grow into serious problems.

Public Life

You and your partner may think you're Dick and Jane, but to your friends and the rest of the outside world you're DickandJane. That means your actions no longer affect just you. Now, they have repercussions for your Other as well. You are DickandJane when you're out running errands, when you're out on the town with friends, when you go to parties and events, when you have friends over to your house, when you go on vacation, and when you attend those special occasions that define milestones in life.

This book's second section, "The Two of You in the World," explores this new reality in depth. In this section, I discuss how your friends can have a major impact on your relationship and how, when you become involved with another person, you also become part of a new family—with all the added personal dynamics this implies. This section will also look at that *other* SO constantly competing for your attention—your work. Your job probably consumes the majority of your waking time; if it consumes you, too, then it may consume your relationship as well.

The Couple You Want To Be

◆

After reading thousands of comments from our couples survey respondents, one overriding message comes through: Successful couples work at their relationships and never take them for granted. They think carefully about the actions they take with their partners. By thinking ahead—by using the principles of etiquette, in other words, whether you realize it or not—you'll avoid many of the potholes on the road of your shared life, and the ones you do hit will be smaller and more manageable. The result is that you are both free to be the individuals you are, and to build the couple you want to be.

PART ONE

THE TWO OF YOU
TOGETHER

1

ETIQUETTE—THE PATHWAY TO A BETTER RELATIONSHIP

THE ELEGANT MANHATTAN RESTAURANT WAS PACKED, WITH the tables so close together that the couple next to my wife and me might as well have been sitting at our table. As a result, it was impossible not to notice what was occurring between them.

A minute or so after the couple sat down, the waiter brought them menus. I noticed, to my puzzlement, that the woman was talking—but not to her husband. Then I realized she had her cell phone to her ear and was conversing with a friend. One by one, she read off each item on the menu, then discussed it at length with her unseen pal. Meanwhile, her husband sat there with his head buried in his menu.

That was bad enough. What happened when the main course arrived, however, was truly astonishing. In the middle of eating, the woman *again* took out her cell phone, called the same friend, and launched into a long discussion about how good the food was—leaving her husband to eat his own entrée in silent isolation. This time, I could tell that he was getting frustrated and annoyed.

You think etiquette doesn't matter when you're part of a couple? Besides leaving an unfavorable impression on everyone around her in the restaurant, the woman's rude behavior turned what should have been a lovely, shared experience for that couple into a serious disappointment on the husband's part. That evening, his wife's lack of etiquette directly affected their relationship—and not in a good way.

So, What Exactly Is Etiquette?

When I first took on the role of spokesperson for the Emily Post Institute, I read all of my great-grandmother Emily Post's books and interviews in an effort to find out what she truly thought about etiquette. I was surprised to discover that Emily actually disliked the notion of rules. What she *was* a proponent of was people having a wonderful time together—engaging in spirited, interesting conversations, getting to know each other well, and doing fun, interesting things together.

In my quest, I came across a perfect description of etiquette that my great-grandmother had given to a magazine writer. It captures the essence of Emily's attitude toward etiquette:

> *Whenever two people come together and their behavior affects one another, you have etiquette. Etiquette is not some rigid code of manners, it's simply how persons' lives touch one another.*

That's it: no manners, no rules—just behavior, and how it affects relationships. Or, to put it another way: The more people's

---

ETIQUETTE IMPERATIVE

Forget About "Which Fork?"

———◆———

After her landmark book *Etiquette* was published in 1922, Emily began receiving letters from people across the United States, asking her every question imaginable about etiquette. One day she received over a dozen letters, all asking which fork to use at dinner. In frustration, she finally turned to her secretary and said, "Tell them I don't care which fork they use!"

Her comment crystallizes what etiquette is all about:

It really doesn't matter which fork you use;
it matters that you use a fork.

---

lives touch each other, the more important etiquette is to the relationship. And what better example is there of lives being intertwined than that of a couple?

The Three Factors That Affect Every Relationship

A relationship is a pretty amorphous thing—tough to grab hold of. If someone simply told you to go out and start doing a better job in your relationship with your SO, you'd probably look at them as if they were nuts.

But what if I were to tell you instead that just by improving your specific, day-to-day *actions, appearance,* and *words,* you can materially affect your relationship with your spouse, partner, or boyfriend/girlfriend for the better—starting immediately? That sounds a lot more doable, doesn't it?

Can it really be that simple? To find out, let's take a closer look at what can happen when things go wrong in these three areas.

Actions. When the woman sitting next to us at dinner picked up her cell phone to call a friend, she instantly cut her connection with the very person she was supposed to be sharing that moment with—her husband. Her behavior shouted, "I'm not considerate of your situation, I don't respect your feelings, and I don't value our time together." Her thoughtless actions spoke volumes.

Appearance. Even if you don't do or say anything, your clothes and grooming send a clear message about what you're thinking and feeling. Another night, while eating in that same Manhattan restaurant, I glanced up to see a couple arriving for dinner. She was dressed very nicely in a skirt and blouse, with a scarf providing an accent of color and style. Her hair was washed and attractively styled—she really looked good. He, on the other hand, was wearing rumpled jeans and a black T-shirt with an inane slogan on the back. "They've *got* to be married to each other," I said to my wife. "Otherwise, there's no way she'd be out with him."

Words. Misunderstandings come from poor word choice as much as anything else. One of the simplest words to misuse is the *we*. The ubiquitous *we* often really means *you*—"Should we call and check the time for the performance tonight?" translates into "Why don't you call and check the time for the performance tonight?"

I can just hear it now: "What do you mean, 'we'? If you want to ask me to do something, just ask!" Better yet, why not simply offer to make the call yourself?

As these examples show, whenever a questionable action, appearance, or word rears its head, it abruptly shifts the focus from whatever activity you are engaged in—whether it's an intimate dinner together, a public event, or an important discussion—to the question "Why is he doing (or looking like) (or saying) that?" When this happens, reversing course and returning the focus to where you want can be difficult.

Out of the Mouths of Babes (So to Speak)

My daughter, Lizzie, is the latest Post family member to work at the Emily Post Institute. Recently, she made the following observation: "Alone, the principles aren't enough," she told me. "When people are in a relationship, everything can't hinge on consideration, respect, and honesty. To make the principles work, people also have to *communicate*, they have to be willing to *compromise*, and they have to *commit* to each other and the relationship."

One of my golfing buddies understands this perfectly. John loves golf. But he also loves his wife and daughter. More than once, I've run into John during the week. "Want to play this weekend?" I ask.

"Sure," says John. "How about Saturday?"

"I can't. I've already got a game. What about Sunday?" I reply.

"Can't," John says, emphatically shaking his head. "That's our family day. I don't play on Sundays."

John and Linda have worked it out. She knows that golf is important to him and she doesn't begrudge him his Saturday-morning game. In fact, she wants him to enjoy it to the fullest. But she also knows he's committed to spending his Sundays with her and their daughter. They've made a pact they both agree on: They've communicated and arrived at a compromise; and by honoring that compromise, they reinforce their commitment to each other.

The Three Principles of Etiquette

Etiquette is governed by three principles: *consideration*, *respect*, and *honesty*. These provide the framework for defining every manner or "correct" behavior that has ever been formulated. These principles are timeless, transcending cultural and socioeconomic boundaries. They apply equally to all ages and all types of relationships—including your closest relationship.

Consideration

Consideration is understanding how other people are affected by whatever is taking place. To be considerate is to show empathy for those around you. Consideration, above all, requires thinking before acting. In order to consider the effect of your actions, appearance, and words on your SO, you'll ask yourself, "How's he going to feel or react if I do that?" It's when you just blindly go ahead and do something without thinking that you're *not* showing consideration—and stuff is likely to hit the fan.

Case in point: Stuart really is a pretty thoughtful guy. That's why, when his mother called recently and sounded a little lonely, he told her to come on over. His wife Rachel had just returned from a workout and was upstairs taking a much-needed break when Stuart came into the bedroom to inform her that his mother was coming over. He'd taken the bull by the horns and done the kind, considerate thing. He was pretty proud of himself.

"You did *what?*" Rachel said. "The house is a mess, I just got back from the gym—*I'm* a mess! And you invite your mother over without even asking me first? What were you thinking?"

We've all heard those fateful words: "What were you thinking?" Answer: You weren't. You just acted, and your actions came from a pretty reasonable place. The problem for Stuart is that their home belongs to both of them. Rachel has a say in what happens there. Moreover, like it or not, Rachel views the condition of their home—not to mention her own appearance—as a direct reflection on her.

Instead of issuing a blind invitation, a considerate Stuart could have said to his mother, "Hey, Mom, can you hold for a second? I want to check something with Rachel."

"Ask her to come by this afternoon," Rachel could have offered. The house would have been tidied up by then, and Rachel would have had time for a shower. That small measure of consideration could have yielded a solution that would have worked for both mother and SO—and a good time would have been had by all.

Respect

Respect is recognizing that how you interact with another person affects your relationship with that person—and then choosing to take actions that will build and enhance the relationship, as opposed to injuring it. Respect helps us decide *how to choose* to act toward others.

One of the most deeply touching displays of respect I ever saw took place in a popular steakhouse in Minneapolis. I was sitting in the bar area when a couple entered and sat down at the table next to me. They were young, maybe thirty at most. Shortly after sitting down, the woman started rummaging through her pocketbook. She searched inside it for a minute or so, becoming more and more agitated. Finally she looked up at her husband and said, "They're not here. I must have left them on the table by the door."

Her husband looked at her and smiled. "Not a problem," he said, starting to stand up. "You order, and I'll go back and get them." And with that he was gone.

She looked at me sheepishly, knowing I'd observed the entire scene. "I forgot the theater tickets," she explained. "I can't believe they're not in my pocketbook. I can see them sitting on that table now. Fortunately, we only live five minutes from here."

"I apologize for watching," I replied, "but what just happened here is one of the most amazing things I've ever seen."

"What do you mean?"

"Your husband didn't make a single snide comment," I said. "Not even a sigh of frustration. He's one in a million."

When I mentioned my amazement to the husband after he'd returned, he told me he couldn't imagine reacting any other way. Now, that's a good relationship—and that's respect.

Getting Even—The Antithesis of Respect

◆

Terry loves driving with the windows open in the car, but his fiancée Jenny can't stand it. Jenny wants them closed. And so Terry has cleverly evolved his own form of retribution: He simply won't play the music she wants on the radio as long as the windows stay closed.

So now they're even.

It doesn't take a genius, however, to see where this kind of tit-for-tat is headed. At best, it leads to mutual resentment and tension; at worst, it can erupt into a major argument.

Because you know your SO particularly well, you know exactly which buttons to push, and how to push them for maximum effect. It's easy to push them when you aren't thinking—you just do it. But the next time you find yourself reaching to push one of your SO's buttons because you're annoyed, frustrated, angry, or merely bored, take a break first; then consider the action again in fifteen minutes' time. Ask yourself: How else might I approach this situation to resolve it in a positive way, rather than in a way that simply gets even with my SO for the moment?

It turns out Jenny can't stand having the windows open because the wind plays havoc with her hair. When they are going out, she doesn't want all her efforts to be ruined before she gets there. Understanding this, Terry and Jenny come to an agreement that if they are just driving around doing errands, the windows can be open and she'll wear a hat or tie her hair back. But when they're going out for the evening or to an event, the windows stay closed. Suddenly, the music selection ceases being an issue, too.

Honesty

Honesty is being truthful, not deceptive. There is a very important difference, too, between benevolent honesty and honesty that is cutting or unkind: "I have a problem with that" is a very different thing from "That's a stupid thing to say."

Honesty ensures that we act sincerely. Sincerity matters. Have you ever listened to a politician speaking and noticed that while he sounds ever so convincing, on some level you simply don't believe him?

It's easy to pick on political figures, of course. But what about ourselves?

"Honey," Steve calls to Jennifer, "it's six o'clock, and the event starts at seven. So we should leave at six-thirty. Are you okay with that?"

"Sure," he hears coming sweetly from the bathroom.

Yeah, right. Jennifer won't be ready until 7:00 at the earliest—and Steve knows it. It's nothing new. In fact, Steve would tell you he's used to it. But their relationship suffers a little bit whenever it happens. Over time, habitually inconsiderate or insincere actions sap the energy from a relationship.

For a relationship to thrive, actions, appearance and words need to be grounded in sincerity. The minute you are insincere, your SO will see right through you. And then you'll have to strive that much harder to recover.

When Manners Are No Guide

As we've seen, etiquette is about more than manners. In fact, manners are really just guidelines for specific situations: They tell you what to do, and what to expect others to do—that's all. You come over to greet me, I stand up. This is what manners call for.

The trouble is, manners don't exist for every situation—especially in a close relationship—and manners also change over time. So how do you know what to do when prescribed manners don't cover the situation you happen to be facing? Simple: Go back to the three key principles of etiquette.

For example, let's say Tom and Jane are sitting with some friends in a booth at a restaurant when Jane's parents approach. Tom is trapped in his seat. Practicing *consideration,* Tom thinks to himself, *I know I'm supposed to stand up, but it's going to be difficult unless everybody slides out. What are her parents going to think of me—in fact, what's Jane going to think—if I don't stand up?*

Next, Tom shows *respect* for Jane by considering how his possible actions, words, or appearance might affect Jane. *I can't stand, but if I don't, that might look bad. I can try to climb over everybody, or have the others get up so I can get out, but that's going to be pretty awkward. How important is it that I stand? Should I say something about the cramped booth, or is my situation obvious?*

Finally, Tom has to be *honest.* Honesty points him to the best course of action. Tom thinks, *Nope, I can't stand, but I'll mention something, so at least Jane and her parents know I was thinking about it.* So, as he extends his hand to shake, Tom says, "Please excuse me, Mr. and Mrs. Smith. It's tough getting up—this booth is so cramped—but I'm really pleased to meet you."

Problem solved. Tom didn't need to memorize a set of manners or rules to know what to do: Being considerate, respectful, and honest guided him to the right solution.

Not Thinking Versus Thinking

Whenever I talk with people about etiquette, I am asked, "What's the problem with people? Are they just plain rude?"

I've met too many nice people to believe that most people are inherently nasty, or that they do rude things intentionally. When nice people do wrongheaded things, what is really going on is that they're acting before they think. And that will get *anyone* in trouble with friends, coworkers, or their SO.

Think about it (no pun intended): How many times have you done or said something without really reflecting on it, only to have your SO (or coworker or a friend or your mother) respond, "What on *earth* were you thinking?"

When that's me, my reaction invariably is, "I don't know. . . . I can't believe I did that."

Classic example: John and Mary are going to a party. John gets dressed: corduroys, polo shirt, and a nice sweater. He plops on the sofa to catch a few minutes of the game before they head out, feeling a glow of self-satisfaction at being ready on time for a change. A few minutes later, Jane comes into the living room—dressed in a drop-dead gorgeous cocktail dress and heels. One look at Mary, and John knows he's blown it. One look at John, and Mary says, "John, you're going to have to change into some better clothes. This is Bob and Sarah's twenty-fifth anniversary party. Everyone will be dressed up. I'm not sure what you were thinking."

There it is again. The fact is, John wasn't thinking. He dressed exactly as he usually does when their friends get together. As soon as his wife spoke, he knew she was right, and he promptly went upstairs to change. Had he been thinking instead of just doing, he would have realized what kind of party it was—or he would have thought to ask.

People like to sing the praises of intuitive behavior. But "acting intuitively" frequently turns out to mean "not bothering to think about how my actions are affecting anyone else."

Intuitive: Dirty clothes spread across the floor.
Thinking: Dirty clothes in the hamper.

Intuitive: Not bothering to check the gas gauge.
Thinking: Checking the gas and filling the tank when necessary.

Intuitive: Drilling a hole in the wall to hang a new lamp.
Thinking: Vacuuming the dust made by drilling the hole.

Intuitive: Not noticing the toilet paper roll is almost finished. *Thinking:* Replacing the near-empty roll with a new one.

Intuitive: Reading in bed when you can't fall asleep.
Thinking: Buying a reading light that won't disturb your SO.

2

ETIQUETTE BEHIND CLOSED DOORS

EVERYTHING SEEMS SO PERFECT BETWEEN BOB AND ALICE. They're great together—the couple everyone wants to spend time with. Out in public, the two of them are in complete harmony. In fact, their friends don't even think of them as Bob and Alice; to everyone they know, they're BobandAlice.

But what about when they're home alone, out of the public eye? No matter how in synch they appear in public, behind closed doors they are Bob and Alice—two individuals living together, each with his or her own views about how the relationship should work. Each has uniquely personal likes and dislikes, and each has habits that can get under the other person's skin.

Somehow, they have to work through all their idiosyncrasies, and figure out how to juxtapose their own individual needs and desires with their partner's. They have to learn to share—their things, their space, and their daily lives—on the most intimate level.

Fortunately, an invaluable tool stands ready to help them accomplish this rather daunting task: Etiquette provides a way for Bob and Alice to navigate their lives together and build a successful relationship—both as Bob and Alice and as BobandAlice.

Letting Etiquette Guide the Way

No matter how smooth their road may appear to be, like all couples, Bob and Alice still run into their share of potholes along the way. How they choose to respond to each new challenge that arises will directly affect their relationship. Etiquette illuminates the road for them, allowing them to navigate these rough stretches. It does this by helping them make conscious, respectful choices that will nurture and sustain their relationship.

Case in point: I had a meeting with my editor in New York City to discuss this book. What should we title it? What about the subtitle? We started having some fun with all kinds of possibilities. As we talked about consideration in a relationship, suddenly she whipped out a piece of paper and jotted down a thought: "Was it good for you, too?" she wrote. We laughed. Maybe that's a title for one of the chapters, we agreed.

When the meeting ended, I rounded up all our materials, including the note. "Be careful with that," she chuckled. "You wouldn't want your wife to find it—at least not without first getting a chance to explain how you happened to have it. It could cause some trouble."

Now, one of the things I tell people all the time about honesty (the third principle of etiquette) is that if you are less than honest—if you tell even a white lie—you are likely to get nailed by Murphy's Law. You know the one I mean: If anything can go wrong, it will.

If I didn't believe in Murphy's before, I sure do now. About a week after coming home, I got a call at the office from my wife. "It's a good thing I'm not the jealous, suspicious type," she said to me. I knew exactly what she was talking about.

"You found it," I said in my most innocent voice.

I went on to explain that my editor and I had actually discussed this possibility, and that I'd assured her it would *never* happen.

Had my wife spent two or three days cogitating on the message, she could easily have dreamed up all sorts of nefarious deeds that I might be guilty of. Imagination might have run rampant, had not communication and respect taken the upper hand. Because my wife felt no hesitation in immediately communicating her discovery, and because I welcomed and respected her inquiry, the incident went nowhere (though I did undergo some well-deserved ribbing).

Making It Work—Just the Two of You

Think about the couples you know who not only have a great relationship when they're out together in public but also have a great relationship when they're alone. They're the couples who are naturally comfortable with each other, who you know are going to make it for the long haul. It looks so easy for them, doesn't it? But I guarantee you—it isn't. You can be quite sure these couples work at their relationship constantly, behind the scenes, when they're among their families, when they're with their friends, even when they're at their jobs. And they don't just focus on the big issues: By using etiquette to smooth over the little rough patches that crop up, successful couples keep small misunderstandings from festering and turning into a major problem.

Here are some specific ways that etiquette can enhance the way you and your SO interact in your daily life.

================ ETIQUETTE IMPERATIVE ================

Don't Be Fooled

◆

The successful couple makes their relationship look so easy. The fact is, nothing is as easy as it looks. It takes work to make a relationship successful.

Resolving Arguments

I was sitting in Dulles International Airport in Washington, D.C. one evening, waiting to board a flight to Charlotte, North Carolina. The good news was that the plane really was going to depart that night. The bad news was, the flight would be leaving two hours late. As I sat in the uncomfortable waiting area, I suddenly realized that I was hearing one side of an argument. A woman nearby was having a loud argument with her SO on her cell phone. Looking around the waiting area, I could see everyone else had become aware of it, too. Here's how the argument went:

"The flight won't get in until midnight."

[pause for response]

"Hey, I can't help it. The flight is delayed. You'll have to pick me up then."

[pause for response]

"*!#%&*! I know it's late! How do you think I feel?"

[pause for response]

"*&%#@! That's not fair. You're never there for me!"

This went on for five more minutes, with no resolution in sight. I had moved by then, and so had most of the other people within earshot. Even being subjected to just the one side of this conversation made us all far too uncomfortable to stay anywhere near her. I don't know what the final outcome was, but I sure wouldn't have wanted to be her SO when he picked her up that night.

The fact is, we're all going to have arguments. I argue with my wife, hard as it may be for me to admit it. My parents argue. I've seen my good friends argue. And I've seen strangers disagree in all kinds of public places. The reality is you're not going to avoid ever having an argument—that's patently impossible. The secret is to learn how to settle the argument in a way that will preserve, or even strengthen, your relationship.

To this end, here are five keys to keep in mind the next time you find yourself mired in an argument, whether it be over something large or small:

1. **Try to stay calm.**

   ✦ It's harder to argue with a person who is calm.

   ✦ Staying calm helps to keep the argument focused on the dispute.

   ✦ Taking some time to get calm can provide an emotional break, giving you a moment to reflect on what's going on.

2. **Use the words *I feel* when describing your emotional state.** "When you watch television without offering to help me clean up, I feel that I'm an unequal partner in this relationship" instead of "You never help out around here!"

3. **Keep the interaction focused on the specific dispute at hand.**

   ✦ Don't stray into old arguments.

   ✦ Don't let your statements deteriorate into personal comments and accusations. These are much more hurtful and take much longer to heal.

4. **Listen.** You may think you are listening, but are you really? Listening means hearing and understanding what a person is saying. In the heat of an argument, it's incredibly easy simply not to hear the other person—sort of like the characters in

Peanuts who hear only "Blah, blah, blah" when an adult speaks. In fact, when you really stop to listen to what your SO is saying, you may discover that—horror of horrors—he actually has a good point that you hadn't considered.

5. **Try to move the argument to a speedy resolution.** Remember that later in the evening, you have to go bed with the person you're arguing with. If you've already come to a resolution and started the healing process, that moment will be a lot more pleasant than if the wound is still open, raw, and uncared-for.

FLASHPOINT

When You're One-on-One

One recently married couple I know told me that on their honeymoon they'd experienced *three* arguments. *No surprise there*, I thought to myself. Take any two people who have just started living together for the first time, put them in a situation where they have to look at each other twenty-four hours a day, seven days a week, and you have a formula for arguments. It doesn't have to be a honeymoon, either, and the flare-ups aren't really anyone's fault: Any time a couple (or other small group) is put into an isolated environment, their overdependence on each other tends to make each person more irritable and sensitive to perceived slights—a psychological phenomenon known as lifeboat syndrome.

Here are some tips for keeping everything smooth and easy in a new setting when it's just the two of you:

✦ Make time to talk.
✦ Ask for your SO's opinion first.
✦ Look for common ground in your discussions.
✦ Be willing to try new things.
✦ Be open to meeting new people.

The Art of Compromising

Compromises come in big and small packages. Small compromise: Pearl wants to watch a chick flick tonight. Harry wants to watch an adventure movie. "Tell you what," Pearl offers. "Tonight, we watch my movie. Tomorrow, I'll pick up some popcorn and beer, and we'll watch your movie. Deal?"

Pearl and Harry managed to negotiate a nice fifty-fifty split. But compromises can't always be so fair to both parties. Here's an example of a big compromise, where the split was clearly lopsided: The husband of a close friend of mine took a job in San Francisco a while back. The couple was living in New York at the time. They had just moved into a spacious new home, and she loved her life there. Selling their house, changing their kid's school, having to make new friends in a place far away from everything she had known—that was a *big* compromise.

The compromise my friend made in moving to San Francisco could never have happened without a firm commitment on her part to sharing her life with her husband. Being committed means you're in this partnership for the long haul, and you're willing to work for it, fight for it, struggle for it . . . because the rewards of its success are worth every ounce of effort.

I visited my friend and her husband in San Francisco recently. Watching them happily building their new life together, I could see that their strong relationship was the reward they'd reaped from their commitment to each other and from her willingness to compromise in order to make his job opportunity possible. And of course, none of it, absolutely none of it, could have happened if they hadn't been able to communicate honestly with each other.

Be Willing to Compromise

◆

The flexible give-and-take of a good relationship depends on the willingness of both parties to compromise.

The Power of a Compliment

Coupling the words "I love you" with a compliment is like pouring rich chocolate syrup on top of chocolate ice cream—absolutely unnecessary, but oh, so delicious. The other night, my wife tried sending me a text message on her cell phone from the other side of the living room. I got it. It read "test message." I looked across the room at her and said, "You could have made it a little more fun than that." I hit Reply on my phone, typed in "Your lips are like fine red wine," and sent it to her. A few seconds later she started laughing. The best part was, her laugh was one of real pleasure.

"You're so much fun to be with" . . . "That meal you cooked was fantastic" . . . "I really enjoyed hearing your take on that new movie." Little compliments like these will add zest, fun, and a welcome touch of appreciation to any expressions of love.

Communicating Your Needs

Have you ever really needed a hug or some other form of reassurance, only to have your SO miss all the signs? When your partner seems oblivious to your emotional or physical needs, it's natural to feel frustrated or to tuck the incident away in your repository of resentments under "things I needed that my SO didn't provide." But these responses are dangerous: Bottled-up frustration will inevitably take you down a path toward venting your building anger at your SO's insensitivity; deposits to your resentment repository will pile up, too, until one day you'll draw on them. And when you do, the proverbial you-know-what is going to hit the fan.

The worst thing about these types of responses is that your SO ends up wondering what on earth he or she did to deserve your anger.

Instead, be as considerate of your partner's perceptions as you would of anyone you were trying to get assistance from. Remember, your SO is not a mind reader. You need some attention, but your partner isn't picking up on your subtle signals? It's time to communicate—not about your SO's insensitivity but about your need: "Honey, could you give me a hug?" or "What a rough day—can we just sit for a few minutes together and talk?"

Giving (and Accepting) Criticism

Criticism tinged with moral judgment and a holier-than-thou attitude doesn't work. Criticism, on the other hand, that seeks to change a behavior without denigrating the offender has a much better chance of succeeding.

Right way: "Bob, what happened? You were out the door before we could talk about plans for today. In the future, could you just check with me before you go?"

Wrong way: "Where did you go? Didn't it even occur to you for one second that maybe we should check each other's plans before you selfishly bolted and left me holding the bag? Don't you ever think about anyone besides yourself?"

Asked the right way, the respectful partner—understanding that his or her action has caused a problem but not feeling personally under attack—gracefully acknowledges this fact and promises to make a sincere effort to act more considerately in the future.

3

COMMUNICATION—
THE GLUE THAT HOLDS
A RELATIONSHIP TOGETHER

LOOKING THROUGH THE RESPONSES TO OUR RECENT COUPLES survey, it becomes very evident that the most successful relationships are all grounded in a continuous effort to communicate. These people told us that they communicate wordlessly with each other by holding hands, giving a hug, offering a back rub. They write notes to each other and send e-mails and text messages. And they talk—first thing in the morning, while out driving or doing chores, over meals, sitting together in the evening, even as they climb into bed at the end of the day.

Think about previous relationships you've been in and why they ended. Undoubtedly, one of the reasons was unsatisfactory communication. Now think of the relationship you are in. As long as you are both communicating with each other clearly and openly about whatever is on your mind, then you have the opportunity to make the relationship work. Stop communicating, on the other hand, and you may very well discover the relationship is heading for trouble.

While channel surfing the other night, I caught a scrap of movie dialogue that hit this nail of failed communication squarely on the head. Two men were vying for the hand of a woman. She clearly liked one more than the other. The only trouble was, the guy she wanted never dropped a single word to let her know how he really felt. Finally she accepted the other man's offer, leaving her true love to bemoan his fate to a friend. The dialogue with the friend went something like this:

True Love: "It's over. It's done. She's marrying him, and there's nothing I can do about it."

Friend: "How do you know? Have you tried? Does she know you love her?"

True Love: "Know? Of course she knows."

Friend: "How does she know? Did you tell her? Did you say the words?"

True Love: "Yes. Well . . . not exactly. But she knows. She wants to marry him. It's obvious."

Friend: "Did you ask her to marry you?"

True Love: "No."

Suddenly, the lightbulb going off in his head, True Love jumps from his chair and announces that by gosh, he's going to let her know how he feels.

I returned to my channel surfing—point taken.

ETIQUETTE IMPERATIVE

No Communication, No Relationship

◆

Without communication, you have no relationship.

The Importance of Saying "I Love You"

I've been married for thirty-two years, and it's easy for me and my wife simply to assume that we love each other. I do all kinds of things to support her and show her that I care, and she probably does even more of these sorts of things for me. Still, no matter how many things she does to show her love for me, it sure is nice to hear her express her love in words every now and then—and I suspect that she appreciates it when I do the same.

Our couples survey respondents told us in overwhelming numbers that saying the words *I love you* frequently was a key element in their relationships. People commented glowingly about how their SO . . .

> "*. . . never leaves the house without saying 'I love you.' "*

> "*. . . makes contact with me at least once a day no matter where he might be just to let me know he loves me.*"

> "*. . . says, 'I love you,' even when he is at work with people around.*"

> "*. . . tells me he loves me every day (we've been married for thirty-four years).*"

> "*. . . tells me he loves me and I am beautiful.*"

"We point to each other," wrote one respondent, "and that means 'I love you.' It works well across a crowded room!" Another replied, "We say, 'I love you,' before hanging up the phone—every conversation."

Saying "I love you" before leaving for work in the morning is a no-brainer. But the survey comments we got opened up whole vistas of different ways these simple words can bring a moment of sunshine to a hectic day. Personally, I think one of the hardest places to say "I love you" is at work. You know what I mean: A coworker's voice suddenly gets real low, and then a quick, mumbled "I love you" spills out. Then there's the non-I-love-you "I love you": "Yeah, you, too." This approach has the dubious double honor of being unsatisfying to the person on the other end of the line and of failing to fool anyone else who may overhear.

While there's no need to shout the words out, there is nothing intrinsically wrong with letting other people—including your coworkers—hear you say "I love you" to your SO over the telephone. You don't have to be dramatic or self-conscious about it; just be natural and real. Contrary to what you might imagine, your coworkers are very likely to think more positively of you for your ability to express yourself so naturally.

You can spice up a compliment or a thank-you by including an "I love you" with it. The next time your SO brings you coffee in the morning, for example, instead of simply saying, "Thank you, honey, this is great," you might try: "Coffee in bed? You are so sweet. Thank you, honey. I love you."

The message "I love you" can also take nonverbal forms such as winks or subtle pointing gestures. Every now and then, I'll see someone lift one eyebrow toward their SO, and a sly smile will then cross both their faces. Cute. Frankly, I've stood in front of a mirror and struggled in vain to learn to lift one eyebrow. I just can't seem to get it.

Beyond "I Love You": The Etiquette of Effective Communication

Saying "I love you" is a great start—but if that was all that was involved in communicating effectively with your SO, this book wouldn't have a whole chapter devoted to the subject. To have a successful relationship, it's also critically important that you and your partner communicate your intentions, ideas, and needs to each other in a clear and consistent way.

Don't Assume Your SO Just Knows

◆

If your SO isn't picking up on your unspoken signals, don't blame him for not being sensitive enough. Instead, tell your partner, in spoken words, what it is you need. Tell him gently. But say it.

If I simply hopped into my car on a Saturday morning and took off without a word to anybody, one thing I know for sure is that my wife would *not* be waiting to greet me with open arms when I returned. Instead, she would rightfully be stewing over the things she never got the chance to discuss with me: Where was I going? Could I get something for her while I was out, and save her a trip? When would I be coming back? What about the path that needs shoveling, or the grass that needs mowing, or the other chores still waiting to be done? And what about her plans to meet her friends for lunch—just who did I think was going to stay home and watch the kids?

When you invite friends or colleagues to your home for a dinner party, you send out invitations several weeks beforehand. That way, your guests have time to make arrangements for attending. Similarly, if you plan to be out of the office for a day, you let your manager and colleagues know several days in advance. Doing these things shows consideration and respect for others and enhances your relationships with them.

These same principles hold when communicating with your SO. Even if you're just running into town on a quick errand, letting your partner know what you're planning to do ahead of time is always the considerate thing to do. By giving your SO plenty of time to react and respond to your plans in a measured way (and being open to his or her responses), you also take the pressure off both of you—and your relationship. And if your plans should change for some reason, keep your SO apprised of that, too.

Communication is also essential whenever any idea or issue arises that affects the two of you. Everybody likes it when one partner takes the initiative to make something happen. But to be successful, that initiative needs to be coupled with appropriate communication, so that the other partner isn't left in the lurch. You have an idea for a trip or a night out on the town or a new look for the living room? Discuss it. Friends have invited you over to their place for dinner at the last minute? Instead of announcing the news to your SO at 5:00 P.M.—"By the way, honey, John and I were talking, and we made plans for us to go over there for dinner tonight; we should probably get there at 6:00"—politely tell your friend that you'll need to check with your partner first, and that if necessary you'll take a rain check. That's communication, that's consideration, and that's building a stronger relationship.

The Art of Listening

Whether you're in a business meeting, socializing at a cocktail party, or just relaxing around the house with your SO, fifty percent of communication is being a good listener. A good listener honors whomever he or she is conversing with by being focused on what's being said, by showing interest through body language, and by offering appropriate responses, questions, and comments.

One of the easiest forms of good listening occurs at the dinner table. By leaning forward and putting your elbows on the edge of the table, your body demonstrates your interest in what is being said. Contrary to what many people think, in her 1922 book *Etiquette* (page 585), Emily Post herself approved of people resting their elbows on the table—both to make conversation easier and also, interestingly, because she felt the posture was more flattering to women:

*Elbows are universally seen on tables in restaurants,
especially when people are lunching or dining at a small
table of two or four, and it is impossible to make oneself
heard above the music by one's table companions, and at the
same time not be heard at other tables nearby, without
leaning far forward. And in leaning forward, a woman's
figure makes a more graceful outline supported on her
elbows than doubled forward over her hands in her lap as
though in pain!*

You and your SO can enjoy honing your dinner-table conversation skills together, both in your own dining room and
when eating out at a restaurant. As you polish your listening
skills, here are a few other tips to keep in mind:

✦ **Make eye contact.** If you failed to make eye contact
while out on a first date, you know you wouldn't get a
second one.

✦ **Don't interrupt.** This is a blatant put-down of the
person you're conversing with. It's wonderful to be
bursting with things to say—but wait your turn before
offering your views.

✦ **Avoid fidgeting.** Jiggling the foot is a classic example of
this. It's a good way to let the other person know you're
either very nervous or very bored.

✦ **Pay attention.** Constantly asking your conversation
partner to repeat him- or herself—"I'm sorry, what did
you say?" "My opinion? About what?"—is another way
of saying, "I'm not really listening to you."

Missed Messages:
Avoiding Communication Snafus

George is watching TV one weekend afternoon when the phone rings. It's Jessica. She and her husband are coming for dinner tomorrow night, and she wants to know what she can bring. George assures Jessica he'll give the message to his wife Rebecca as soon as she gets back from the health club.

Now, two things can go wrong: Either George can forget to give the message to his wife or he can dutifully give her the message but Rebecca can fail to call Jessica back.

Whichever person it is who drops the ball, the problem isn't just his or hers—it's *theirs*. Why? Because the mistake ends up reflecting badly not only on the transgressor but also on the other person involved in the communications slipup. George or Rebecca makes the mistake as an individual, but the opinion that others hold of them as a result reflects on GeorgeandRebecca.

In this particular case, the faux pas isn't a deal breaker. Still, like it or not, these communication breakdowns tend to add up. That's why George and Rebecca are scrupulous about writing down careful messages for each other and posting them conspicuously. They focus on who called, the time and purpose of the call, and whether the caller expects a return call or will phone back. Also, whenever one of them listens to their home voice mail messages, he or she is careful to save all messages intended for the other. By proactively sidestepping little bumps in the road such as these, George and Rebecca maintain a stronger relationship not only between themselves but also with the other people in their lives.

Your Tone of Voice Matters, Too

Just as important as what you say is the tone of voice you use. An undercurrent of anger, an air of distraction, or an edge of sarcasm will influence how your partner hears what you're saying. Even if your tone is simply misinterpreted, this will garble your message at best, and cause a major misunderstanding at worst. If, on the other hand, that anger or distraction creeping into your voice is a sign of how you're really feeling—then it's time to, yes, talk about it.

The importance of tone gets elevated to a whole new level when speaking on the phone. Since your listener can't see you, he or she has no clues to help decode the intent behind your words except for the tone of your voice. Talk with your SO in person, and that twinkle in your eye sends a clear signal that your sarcastic comment was meant as a joke. Use the same sarcasm in a phone conversation, however, and it could lead to a full-scale donnybrook.

Taking a TV Time-Out

I quake at the thought of how watching television has replaced talking with each other in our lives. It comes on first thing in the morning (some TVs even have alarms); we watch it while we're eating; it puts us to sleep at night. If *The West Wing* is on, at least in my house, don't even think about discussing anything. I'll even consider letting the answering machine pick up our calls for that hour.

That's why it's vitally important to set aside times when you and your SO sit down and actually talk together. That means no TV on to distract you—not even in the background with the volume turned down. One of the best ways to do this is to resurrect the evening meal, where you and your SO sit at a table and share dinner and conversation. Another way is to plan a regular activity that gives you talking time together, such as strolling through a park or taking a drive in the country.

Written Communication

Following a fairly heated argument, Susan and Bill retire to separate parts of the house for a while. Later, Susan finds a note on
her pillow: "Meet me in the den," it says. When she opens the
door, she sees the lights turned down, two glasses of wine on the
coffee table, and Bill holding his hand out to her. The healing
has begun.

Instant messages, e-mails, and handwritten notes are all ways
to communicate eloquently without saying a word. Often these
forms of communication say much more about our real message
than spoken words can.

Note Writing—A Lost Art That Speaks Volumes

Notes can be used for much more than patching up a quarrel.
They're also a great way to add spice to your SO's day. The following note-writing tactic is one I like to suggest to my seminar
participants. It never fails to surprise them: The next time you
and your SO go out for a special evening on the town, sit down
the next morning and write your SO a note. Tell him what a
wonderful evening you had, how glad you are to be sharing your
life with him, and how much you love him. Keep it simple—just
a few lines will do. Then put the note in an envelope, address it
to your own home, and send it off by snail mail.

Time and again, our couples survey respondents cited notes
and e-mails as special ways their significant others made them
feel loved and cared for. Here are just a few examples, along with
my own comments in parentheses:

> "He e-mails a simple 'I love you' at some point every
> morning, despite the fact we live together." (One of the best
> uses of e-mail I can think of—it's quick and easy and
> has immediate impact.)

E-mails Are Not Private

◆

The stories are legion—here's how one version goes:

They had a wonderful romantic evening. The next morning she was still glowing, so she decided to write him an e-mail at work. Her affectionate message didn't spare any details in extolling the pleasures of the previous evening. Unfortunately, the e-mail got out somehow. (Frankly, the how doesn't matter.) As a result, her office, her town, and people around the world had the opportunity to enjoy her account of the evening, too.

Here's the bottom line: If you wouldn't post that message on a bulletin board for anyone to read, don't send it in an e-mail.

"He surprises me by leaving me notes in my car." (Notes in unexpected places heighten the impact of the message.)

"He sends me regular e-mails during the day telling me how beautiful I am." (You ratchet the note up a notch when you add a compliment to it.)

"Leaves me love letters every morning before he leaves for work." (Consistency is great—just be aware that once a pattern is set, you'd better not slip up, or your SO will wonder what's going on.)

"Leaves a note saying 'I love you' on the refrigerator if we don't see each other in the morning." (Notes help you connect when you can't be with each other.)

"We both write a lot of notes and hide them in luggage or under pillows." (Written messages don't have to be in plain sight; the act of discovery can be as much fun as the note itself.)

"He'll send me a card out of the blue." (A carefully chosen card will make your message even more memorable.)

"We write love notes on the bathroom mirror." (Who needs notepaper or a personal digital assistant? The bathroom mirror, freshly fallen snow, sand—there are lots of places to write a note.)

I think it's the surprise of the note that makes getting it such a pleasure. In the middle of doing some mundane, everyday task, suddenly you find a written message from someone you care about that makes you feel fantastic. It's such a small, easy thing to do—and it has such powerfully positive benefits for any relationship.

When Communication Breaks Down

◆

There might come a time when, despite all your best intentions, you and your SO find yourselves unable or unwilling to communicate about one or more key issues in your relationship. When this happens, it's time to seek professional help. Marriage counselors, couples therapists, and many members of the clergy are trained in helping couples talk about difficult subjects. Counseling may or may not resolve your issues, but one thing is for sure: Not talking about your problems is a surefire way not to solve them.

4

THE IMPORTANCE OF
NONVERBAL COMMUNICATION

CAROLINE SUDDENLY STARTED LAUGHING. "SEE, HE DID IT again—it must be time for them to go."

We were at a party at Toby and Caroline's house. Nice people. And the party had been terrific. The problem for me was simple: I had been to soirees at Toby and Caroline's before, and I knew what was likely to happen. My wife had developed a habit of being the last person to leave the party. One night, she held out until three in the morning.

I was tired and eager to head home, but I also didn't want to just blurt out, "Time to go!" I simply wanted to let my wife know I was ready to leave—so what did I do? I gently rubbed her back. I thought I was making a subtle sign, the kind of thing a husband does to make his point without drawing attention to it. Little did I know that my "subtle" signal had become so well known that everyone at the party was waiting for it. As a result, it was now having exactly the opposite effect from what I'd intended: Far from being discreet, I had drawn attention to my "let's call it a night" message. (We ended up staying until the end of the party anyway.)

Although it backfired in this case, my rubbing my wife's back was a classic example of the silent messages that couples use to augment their spoken words. Nonverbal communication is built on such simple things: a touch, a squeeze of the hand, a wink or look. Yet surprisingly, they really make a huge difference. These little signals are how we tell someone we love and appreciate them or that we're enjoying their company or, maybe, that it's time to go. They are an essential part of the way a couple communicates with each other.

When we asked our couples survey respondents what special nonverbal communication they used with their SO, we received a variety of unique and intriguing answers. But virtually all of the responses mentioned certain characteristics:

+ Physical signals take on meaning because they are repeated over and over again. Through that repetition, they become more and more important. For example, many people commented on the fact that their SO holds their hand every night as they fall asleep.

+ Meaningful nonverbal communications tend to be conscious moments of expression. With that wink across the room, you briefly but explicitly focus on your SO, making him or her feel special.

+ These physical expressions are special to the couple, sometimes to the point of being private or even secret— the soft touch that no one else even notices (you hope), that creates a little spark between the two of you.

If our survey results are any indication, respondents who engage in regular nonverbal communication are, as a group, quite successful and satisfied in their relationships. The bottom line: Physical communication is a language that successful couples understand. It makes a difference.

ETIQUETTE IMPERATIVE

Power Caresses

———◆———

Make nonverbal messages a part of your life, and use them regularly. Using a physical expression consistently to communicate your feelings will make that moment when you say "I love you" even more real and meaningful.

The Signals We Send

Jeremy and Janet are in the midst of a disagreement. Janet is leaning against the kitchen counter, arms folded across her chest, shoulders hunched, eyes throwing darts. Meanwhile, Jeremy has rolled his eyes to the sky, turned his back on her, and walked out of the room. It's clear what's going on, without anyone having to utter a syllable.

Our body language and other nonverbal signals can communicate our thoughts and feelings just as powerfully as words do. Women seem to have a Ph.D. in nonverbal language. I used to think that men were dunces at it, so I was amazed to see that in our couples survey, female respondent after female respondent talked about how important her male SO's nonverbal communication was in their relationship. Men seem to have perfected two modes in particular, both of which speak volumes to their intended recipient:

✦ **The wink.** Here's what our respondents had to say about this simple gesture:

"A wink doesn't cost anything, and puts a smile on my face."

"He's a musician, and he winks at me from onstage when he's playing."

"When I'm in the middle of a conversation, he winks with a grin, and then walks away."

"I love it when he winks and smiles."

✦ **The look.** The exact details of "the look" may vary from couple to couple—but our survey respondents agree they love it when their SO . . .

". . . gives me that special look."

". . . sends a complimenting glance."

". . . makes eye gestures."

". . . gazes into my eyes."

". . . gives me his 'cute face.' "

". . . really looks at me."

". . . stares into my eyes and smiles."

It's interesting to note that both of these forms of nonverbal communication involve the eyes. When we aren't speaking, we still communicate very effectively with our eyes. So be aware of what they're saying to your SO: Your gaze can show joy, sorrow, love, passion, laughter, wonder, anger, hurt—all without speaking a word.

The Glare, the Sigh, and the Slump

Unfortunately, nonverbal communication isn't always positive. When we're angry, annoyed, or upset, our bodies can give us away even when our mouths are shut. Too many people are famous for their withering glares. Nobody wants to be on the receiving end of such a look—but the truth is, you don't really want to be on the giving end of one, either. It's a pretty brutal form of communication, one that is devoid of any nuance. And once expressed, it's tough to water the message down: "Oops—sorry, honey. I didn't really *mean* to look like I'm livid. Just ignore my scowl, and listen to my words." Yeah, right.

Negative Signals—Worth Sending?

I know that I can shrug my shoulders and roll my eyes in a way that very effectively says, "You've got to be kidding. Take the garbage out *now*? There's one minute left, and the score is tied 2–2."

It's so easy, too: Pursed lips and squinted eyes can express frustration better than a thousand words, while standing with folded arms and an impatiently tapping foot or pacing back and forth are wonderfully direct ways to let my wife know we're running late.

But these signals are rarely well received. Often, we send them on the spur of the moment, without truly thinking about how they might be taken or even about how valid the message actually is.

Is it so difficult to say, "Let me just watch the end of this game and then I'll take out the garbage"? Does it really matter if you're going to arrive at the party five minutes later than you'd wanted? Take a deep breath and relax. Better yet, tell your SO how great she looks when she comes down the stairs. Giving in to exasperation starts the evening with a black cloud over everybody's head. Being patient—and paying a heartfelt compliment on top of it—will have you both feeling good as you go out the door.

While a grouchy sigh or slump may not be as devastating as a skewering glare, these actions can also communicate displeasure all too well. I know this firsthand, because I'm very capable of heaving a frustrated sigh when I'm being selfish and things aren't quite going my way.

The problem with this kind of negative body language is that besides making your SO (and everyone else in the vicinity) uncomfortable, it doesn't do a thing to fix whatever is bothering you. In fact, it makes matters worse, by placing the burden of figuring out what's wrong onto your SO.

Take a simple situation that happens fairly often in our house. When I'm watching television, I want to watch television. Conversation makes it difficult to hear, much less concentrate, on the show. Normally, with just my wife and me in the house, this isn't an issue. But when my two twenty-something daughters are home as well, talk is abundant—and I've been known to let many a deep, heartfelt sigh slip out. Bad Dad.

Gradually I'm learning simply to ask them to talk somewhere else instead. The advent of the digital video recorder, which lets you pause a show at any time and restart it in the same place— has really helped curb the sighing, too. Hey—whatever works.

The glare, the sigh, the slump . . . they *don't* work, and I'm working to eradicate them from my body's vocabulary.

Communicating Through Touch

Touching is a powerful way to communicate without words. Gentle touches and other little physical gestures are a way of expressing empathy, conveying a sense of connectedness with your partner, and soothing frayed nerves. As one of our couples survey respondents put it, "A gentle touch when I am stressed means more to me than a long, drawn-out conversation on why I'm stressed—which usually tends to stress me out even more!"

Touching Matters

———◆———

Americans are not a touch-oriented people, by and large. Think about it: Besides your SO and your children—more likely when they're little—we don't do a whole lot of touching. In fact, if a member of the opposite sex touches you—a squeeze of the shoulder, a light rubbing of your arm or back, or reaching out to take your hand—you may immediately wonder, *What's* that *all about?* And as for members of the same sex, I can't remember the last time one of the guys in my golf foursome touched me except to shake hands and pocket a few bucks. If it was an especially egregious loss on my part, he *might* give me a quick slap on the back, just to rub the hurt in a little more.

With our SO, however, things are different. He or she is very likely the only adult person we touch with any frequency. Touching is a way of communicating with your SO that is uniquely yours to use, and one that can have a very positive effect on your relationship. There are lots of ways to touch—hug, kiss, rub, hold hands, squeeze. Do it gently and with feeling to let your SO know you care. Your partner will think warmly of you for it.

Our respondents told us that touch is both a way to communicate when they are alone together and a means of connecting privately in public. The survey also showed that touching takes many forms:

"A random touch at some point in the day—just a fleeting contact that lets me know he's aware of my presence."

"He reaches over in the car to take my hand for a quick squeeze."

"He places his hand on the small of my back and leads me into a room."

"We play footsie."

"He puts his hand on my back while we're standing next to each other in public."

"She strokes my thumb while holding hands in public, which is our secret code for 'I love you.'"

"He touches me gently to let me know he cares and is concerned."

"When one of us walks past the sofa, the person sitting down usually sticks out a hand for brief contact, or the person passing by touches the other's arm."

For all the wonderful varieties of touching that our respondents reported, however, when it came to describing what contributed to a great relationship, two types of nonverbal communication stood out above all others: holding hands and kissing.

Speaking With Your Hands

I can remember, clear as a mountain stream, the excitement of going to the movies with a girl back when I was a young and innocent teenager. The big question afterward always would be: Did you or didn't you hold her hand? When you were able to hold hands, it felt as if you'd struck gold—*if she didn't pull away.* There was a spark to that physical contact, especially the first time it happened—and that spark never completely disappears.

The gentle pressure of fingers interlocked with fingers still affects me when I hold my wife's hand today. It seems like we hold hands a lot: in the car going to or from a party or event (kind of like being on that date again), watching the evening news on television, sitting in bed in the morning, enjoying a cup of coffee before attacking the day.

I used to think that maybe we were an anomaly, but lately I've discovered that we're not—not by a long shot. Holding hands was by far the most common expression of intimacy cited in our couples survey. Our respondents reported that they held hands walking in the street, at a party, lying in bed, watching TV, and while praying. Here's what some of them said about the importance of this simple gesture:

"He softly squeezes my hand without saying anything." (You don't have to speak to let your SO know you care. That simple gesture conveys the message loud and clear.)

"We hold hands a lot, but I don't know if it's him holding my hand, or me holding his. Our hands just find each other." (Holding hands becomes a natural and important part of the relationship. I know that when my wife and I walk down the street, I feel naked if we're not holding hands.)

"He's always reaching for my hand if I don't reach for his. We hold hands even at home when watching TV." (Holding hands is not a public declaration—"Look, aren't they cute, holding hands after thirty years of marriage"—it's something we do in private, too.)

"She holds my hand when we're praying." (Holding hands can become a part of a bigger, more important moment, by adding a sense of unity that is the essence of being in a relationship.)

"My SO reaches for my hand when we're out in public, or in the middle of the night." (It's amazing how many people wrote about the importance of holding hands in bed when they're going to sleep, in the middle of the night, and on first waking up.)

Holding hands doesn't have to be a passive thing. Squeezes and playing with each other's fingers have a key place in the hand-holding continuum. Several respondents revealed their secret codes—almost a Morse code signaling system, in some cases:

> *"We have a special, secret hand squeeze—he'll squeeze my hand three times quickly, which signifies "I love you," then I respond with two squeezes back, meaning "How much?" He then squeezes longer—then I do four quick squeezes: 'I love you, too.'"*

> *"We squeeze each other three times, usually on the hand or above the knee. The three squeezes are a way of saying, 'I... love... you...' without speaking."*

> *"She squeezes my hand three times to say 'I love you' when she can't tell me out loud, like in church or at a play."*

Hand in hand, you and your SO are a couple. Your joined hands say so to the world and, most important, to your partner.

ETIQUETTE IMPERATIVE

Make the Effort—Hold Hands

◆

Hold hands with your SO. It's the simplest and easiest of gestures, and one of the most powerful. So just do it.

The Language of Kissing

There are many types of kisses. While passionate lip-locks are well suited to the heat of the moment, kisses that are expressions of affection—whether exchanged in private or in public—tend to be tender affirmations of love rather than the precursor to a more intimate moment. A kiss of this sort is a clear way of saying "I love you" to your SO and proclaiming your love for your SO to others.

Kisses come in many shapes and sizes and say many different things. The beauty of a good kiss is that it speaks for itself—the words may follow for emphasis, but often you don't have to say anything. Sometimes that's the best way to communicate.

✦ **Saying "Good morning":** Kisses often become part of a relationship ritual: "He kisses me first thing, each time he sees me during the day" . . . "He never fails to kiss me good morning when he gets up."

✦ **Surreptitious messages:** At social events, a blown kiss done subtly lets you make a connection with your SO without anyone else around you realizing what's going on: "She blows me a kiss or winks at me when we're at a party" . . . "the special way he throws a kiss or wink across the room."

✦ **Affection without a flame:** Forehead kisses seem to be a popular way to show affection without lighting a fire, even when you're alone with your SO: "A gentle forehead kiss" . . . "His kisses on my forehead first thing in the morning are not only tender but also respectful of the fact that neither of us has brushed our teeth yet!"

✦ **Public quick kiss:** You know the one—it says, "It's so much fun to be with you!"

✦ **Soft, gentle kiss:** It's been a tough day. Or you've been feeling a half bubble off plumb all day long. Or you're feeling rushed. Suddenly, you receive a soft, gentle reminder that someone does care and wants you to feel like a million dollars.

✦ **Memory of a flame:** Last night was wonderful. You both are still basking in the glow. The kiss has the memory of the passion in it.

✦ **A compliment:** You've made the effort to dress your best and look your best. When your SO sees you, the first response is a kiss and "You look great tonight."

When Kissing Language Becomes Foul

Jenine was riding on a bus recently. She often rides on buses, so that wasn't unusual. The couple across the aisle was. They were lip-locked. Not a quick, soft kiss. No, this was unbridled, open-mouthed lust. It wasn't "Oh how cute, they must be new lovers." It was down and dirty, and nobody around them enjoyed the spectacle. Jenine tried to move as far away from them as she could, but the bus was crowded and there wasn't much room.

A public display of affection is one thing. Sharing sex with a busload of people is another, and it's not something anyone else should have to be subjected to.

5

THOUGHTFUL GESTURES MAKE ALL THE DIFFERENCE

THOUGHTFUL GESTURES COME IN MANY SHAPES AND FORMS. We can literally reach out and touch someone—giving a hug, for example. Thoughtful gestures can also be actions that ease your partner's burden—walking the dog early in the morning so that he or she can sleep in, washing the dishes when it isn't your turn, or going out of your way to pick up an item at the store that you know your SO needs. They can range from the unexpected (surprising your partner with tickets to a show) to the exotic (fixing your SO a special candlelight bath) to the sensual (giving him or her a massage with scented oil). The form a gesture takes doesn't matter nearly as much as the fact that it allows you and your partner to make a connection.

✦ How about the husband of a nursing mother who makes the effort to ask her what he can bring her to drink while she is breast-feeding? A simple gesture, really, yet one borne of a willingness to think about her situation and what she might like, and then *act* on it.

✦ How about the person who spends extra time talking to her SO's grandmother on the phone?

✦ How about the person in a long-distance relationship who makes a point of stocking up on his SO's favorite snack foods and reading material before each visit?

✦ How about having a beverage and hors d'oeuvres ready and waiting for your SO when you know he or she has had a particularly trying day at work?

✦ How about agreeing to do something you don't really want to do—*and* not complaining, even once?

All of these things require one trait above all: the ability to stand in another person's shoes and sense what will make that person's life easier, happier, or better. Let's take a closer look at some other relatively minor gestures that can make a major difference in your relationship.

Making Coffee for Your Partner

Making coffee for my wife is one of my favorite thoughtful gestures. For some reason, I always thought I was the only person in the world who did this every morning—then I saw the results of the couples survey and discovered how many people do this for their SO on a regular basis. So much for originality!

Some people do it on the weekend. Others do it intermittently. And then there are those folks like me who do it consistently. I'm the first to rise, and my routine is the same every morning: I go downstairs, feed the animals, then make coffee and bring a cup to my wife, who is just waking up.

While I was surprised by the sheer number of people who get coffee for their partners, I was even more taken aback at how incredibly impressed their significant others were by this simple, thoughtful gesture. It's not just about the coffee, either—it's also about the shared moment together drinking it. "Bill brings me coffee and we talk our heads off," wrote one respondent. "It's great."

Over time, this coffee thing can become a ritual—a gesture that is repeated virtually automatically, regardless of the circumstances:

> *"Brings me my coffee every morning (20 years and counting!)"*

> *"Buys me coffee from 7-Eleven and a paper every morning."*

> *"Brings me coffee in bed no matter how she's feeling about me."*

I remember one Saturday morning: I went downstairs fairly early and decided not to take a cup to my slumbering wife because I didn't want to wake her. A short while later, she appeared in the kitchen, where I was sitting absorbed in the paper. "What happened to my cup of coffee?" she asked.

I realized then and there that this gesture had become a ritual—one that I shouldn't mess with. I haven't repeated that mistake. Now, if she's still sleeping, I just leave the coffee on the bedside table and pad quietly away.

Massages and Rubbing

Dogs like to be petted. If my dog's reaction is any indication, being petted must feel fabulous.

We like to be petted, too—having our arms or legs gently rubbed, our sore muscles gently massaged, our feet or hands squeezed, or our back scratched. Like kisses, rubs and massages can lead to all kinds of wonderful things, but they're also great gestures in and of themselves. In fact, I think that if massages always led to intimacy, they would lose their effectiveness as a gesture. A rub becomes an even more enjoyable connection by *not* always allowing it to turn into something more. Let a rub take the stress out of a hard day, soothe the ache of tired feet, or simply make your SO feel good. The focus is on the sensation at that moment. It is the simplicity of the act done for its own sake, without any expectation of more to come later, that makes a rub or massage so pleasurable. Just ask my dog.

Hugs

At social gatherings, it seems as if hugs are now the standard greeting for members of the opposite sex. Then the men shake hands with each other and the women do the hug–air kiss thing. Once in a while, I'll run into a guy who goes for the hug as part of the greeting, but that still seems to be uncommon.

With your SO, however, a hug becomes a more poignant gesture. It's a wonderful physical way for you to share your affection with your partner. Best of all, it can be done in all kinds of places. You can even give a hug as you're walking down a street with one arm around your SO's waist, by simply tightening your arm muscles and giving a little squeeze—which your SO will feel as a hug.

That hug you just gave him or her is what I like to call a private moment in public. There's a special bond forged when you communicate something just to your SO, with no one else knowing about it.

Hugs can also be a surprise, and pleasant surprises in a relationship are definitely nice things. Our couples survey respondents noted the added pleasure of a totally unexpected embrace—"a surprise hug at an unexpected time (from behind, in the middle of a conversation, etc.)" or "sneaks up behind me and squeezes me when I'm cooking, etc."

Finally, hugs, like kisses, simply say "I love you": "gives me a hug and a kiss when he comes home from work, because he missed me," said one respondent.

Flowers

Flowers always have been, and always will be, a surefire way to make your SO feel very warmly toward you. A note to women: Don't forget to give your man flowers from time to time. In fact, I'd always wondered what it would be like if a man received a bouquet of red roses from his SO at work. This gesture, it seemed to me, would create quite an impression on his fellow workers—males and females alike! And then, while I was in the midst of writing this book, one woman told me the following story:

"My cabinetmaker husband's coworkers were mightily impressed when he got a dozen roses from me at work. He enjoyed all the speculation as to what wonderful thing he'd done to deserve the flowers."

Now I know.

A note to men: You get extra points by knowing what your partner's favorite flowers are, rather than simply grabbing the first arrangement you see.

ETIQUETTE IMPERATIVE

Complete an Unfinished Task or Errand

◆

Surprise your SO by doing something the two of you have been meaning to get around to—such as finally getting that terrific print framed, installing the bird feeder in the backyard, or organizing your photo collection.

Giving Your Partner Time Off

Take over all household responsibilities for part of the day, so that your SO can do whatever he or she wants—whether it's shopping, getting a manicure, playing golf, or just lounging in front of the television.

Going Along for the Ride

Accompany your SO even when he or she says, "You don't really have to"—whether it's a quick errand, a visit to the doctor, or ferrying the kids around town.

Be an Active Listener

Be ready and willing to act as a sounding board for your partner, even if it's only about "trivial" stuff like office politics or a recounting of your SO's dreams the previous night. The key is to let your partner know you're listening.

When possible, stop what you are doing so you physically can give your entire attention to your SO. Start by looking your partner in the eye. At appropriate breaks as your partner is talking, interject something more than just, "Un-huh." Ask a question or make a comment that shows you were listening: "Did she

Supporting Your SO in Difficult Times

◆

When things are going well, it's fun to know that a morning cup of coffee, a night out, or a present can help make your SO's life and mood even better and, in turn, make yours better, too. It's easy to make good times great.

Unfortunately, life isn't always sunshine and roses. Everyone has stress in their daily lives, and everyone also endures their share of hard times—difficulties on the job, health problems, sickness or death of a loved one, friends or relatives going through a divorce. When your SO is struggling—either with a particular problem or because he or she is simply feeling stressed out from the daily grind—supporting your SO actively and considerately can make a huge difference in the way the two of you—and your relationship—weather the storm.

Among the many thoughtful gestures that couples survey respondents identified as being important to their relationships, the ones that drew the most glowing praise were gestures of compassion and consideration in rough times—not only because they were much needed but also because of the sincerity and love that they demonstrated:

> *"He recognizes that I need a break now and then from our children."*

> *"She is there to comfort me when I need it."*

> *"He sleeps on the couch when I am having a bad night of pain."*

really say that?" "What did he do next?" "Seems like he's making a mountain out of a molehill."

The questions and comments become even more important if you can't stop what you are doing—like making dinner. Then they're the only tool you have to let your partner know you are listening.

The Power of Sharing

As children, we learn to share things like toys and friends. As we grow older, sharing deepens. We share experiences. We share thoughts. We share feelings.

I'm not saying that *every* detail has to become fodder for your relationship. For example, I don't really want my wife to share with me the specifics of all the gossip that was passed around on one of her girls' nights out (See "Boys'/Girls' Night Out," page 166). But being able to share in the pleasure my wife got from having fun with her friends is terrific. Not only does she tell me what a fun time she had, but I can also see from her mood and attitude how much she enjoyed herself. And that fact makes the night a pleasure for me as well.

6

MILESTONES
AND SPECIAL DAYS

WE ALL LEAD SUCH BUSY LIVES—SO MUCH FASTER-PACED, more complicated, and more informal than those of our parents and grandparents—that taking time out of our packed schedule for traditions and special days can seem like a luxury. Think about it: How many times have you or your SO bought each other a birthday card or a Valentine's Day card *before* the actual day of the event? A more likely scenario is this: It's February 14; you look at the clock and realize that you'll be heading home from work in just a few minutes, and you haven't gotten your SO a card, a gift, flowers . . . *anything.* And so you battle through the snow and freezing rain to the card shop. The rack is almost empty, and the only remaining cards are a sorry lot. You grab the first decent one you see, along with whatever small stuffed animal or heart-shaped box of chocolates happens to be on the store's shelves. Then you head home, celebrating the fact that you've dodged yet another bullet.

Does it really matter whether or not you remember your SO's birthday, the anniversary of your marriage, or any other special day? Here's what our couples survey respondents had to say:

> *"My SO remembers special events from years ago."*

> *"My partner gives me very thoughtfully planned birthdays and special occasions."*

> *"We wish each other happy anniversary on the fourteenth of every month (we were married on September 14)."*

> *"He always remembers my birthday, our anniversary, and Mother's Day."*

Some special days will be unique to the two of you, such as your wedding day or the anniversary of your first date. Others, like Valentine's Day or Mother's Day and Father's Day, are shared with the whole world. The wonderful thing about both types of occasions is that they give you a number of specific opportunities, all plotted out in advance on the calendar, to make a special gesture that will make your SO feel loved, appreciated, and cared for.

Of course, the downside to these special days is that if you *don't* do something to mark them, you risk looking like a heel. The bottom line is clear: One of the simplest and surest ways to improve your relationship is to make a point of consistently remembering and celebrating those special moments that define your relationship.

Celebrating Anniversaries and Birthdays

Neither partner wants to forget either of these special days. Often, you'll hear someone claim they don't really care that their partner forgot their birthday or only notices at the last minute that their anniversary is coming up.

Don't believe it.

These protestations are just a way of making the unacceptable acceptable. And even if someone truly isn't bothered by their partner's oversight, it doesn't mean they wouldn't have been touched and pleased if their SO had made more of an effort to honor them. Being told we're special is always welcome. In fact, chances are good that eventually you'll be at a party or dinner or whatever and you'll hear the aggrieved party make a caustic comment directed at their SO like, "It sure is nice that John remembered their anniversary with such a beautiful diamond pin. I wish Tom would remember our anniversary just *once.*" And so commences round one of what's sure to be a fight once they get home.

One reason why anniversaries and birthdays matter so much is that, while the event itself happens only once, the memory can be relived many times. Make a sincere effort to celebrate these special days, and the memory you create will be something both of you will treasure the rest of your lives.

ETIQUETTE IMPERATIVE

More Than a Day on the Calendar

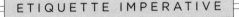

Remembering and honoring special days *does* matter—a lot.

Anniversary Gifts

———◆———

While some anniversaries are clearly bigger occasions than others, every anniversary is an event that should be celebrated. Here's a list of yearly anniversaries and the gift (or gift type) associated with each, with some ideas thrown in for good measure:

1st anniversary: Paper or plastics—books, notepaper, magazine or newspaper subscriptions

2nd anniversary: Calico or cotton—cotton napkins and place mats, cotton throws, tapestries

3rd anniversary: Leather or simulated leather—photo album, leather bag or suitcase

4th anniversary: Silk or synthetic material—silk flowers, silk handkerchiefs or scarves

5th anniversary: Wood—picture frames, hand-painted wooden trays, wicker baskets

6th anniversary: Iron—fireplace tool set, wind chimes

7th anniversary: Copper or wool—copper bowls, pots, or kettle; wool afghan

8th anniversary: Electrical appliances—hand mixer, blender, waffle maker, espresso maker

9th anniversary: Pottery—ceramic vase, platter, pitcher, bowl, set of mugs

10th anniversary: Tin or aluminum—attractive cookie or biscuit tins, mailbox, rustic birdhouse

11th anniversary: Steel—stainless-steel kitchen utensils or bowls

12th anniversary: Linen—damask tablecloth, fine bed linens

13th anniversary: Lace—tablecloth, napkins

14th anniversary: Ivory (now outlawed), antique jewelry

15th anniversary: Crystal or glass—Christmas ornaments, carafe, wineglasses

20th anniversary: China—hand-painted bowl, platter

25th anniversary: Silver—ice bucket, wine bucket, engraved goblet

30th anniversary: Pearls—pearl-handled steak knives

35th anniversary: Coral (now endangered)—brooches, figurines

40th anniversary: Ruby—jewelry

45th anniversary: Sapphire—jewelry

50th anniversary: Gold—jewelry, gold-leaf stationery

55th anniversary: Emerald—jewelry

60th, 70th, and 75th anniversary: Diamond—jewelry

Gifts Matter

I have a good friend who, over the years, has received a number of practical anniversary presents from her husband—a dishwasher, a stove, new wineglasses, luggage. Recently, though, she was sporting a very nice new gold necklace. More important, I haven't seen her beaming like that in a long time.

"Do you see what he got me for our anniversary?" she exclaimed. "He picked it out on his own. I can't believe it!" I couldn't be happier for them: She's ecstatic; he's her hero; and their relationship has gotten a wonderful boost.

That doesn't mean you've got to go for the gold (or the diamonds) on every birthday or anniversary. A modest gift chosen with care—care being the operative element—can be as treasured as an extravagant gesture.

A gift doesn't have to be traditional or, for that matter, a surprise. More than once, my wife and I have agreed ahead of time that a trip abroad, or getting something we really want for our home or yard, would make a great present that year. The key is that we talked about it. Still, I can assure you that if I were to happen to pick up a little something extra—something that wasn't practical, something she wouldn't ordinarily go out and buy for herself—my extra effort would be very much appreciated, and vice versa.

Special Milestones Matter Most of All

All birthdays and anniversaries matter, but the big ones—such as the first, the fifth, the tenth, the twentieth, the twenty-fifth, the fiftieth, and so on—really matter. On these occasions, special care is required.

My mother and I were born almost exactly thirty years apart, with my birthday just three days before hers. Five years ago, to mark milestone birthdays for the two of us, my parents joined my wife and me on a trip to Italy. We celebrated with birthday dinners in a couple of fabulous restaurants, one in Rome and one in Tuscany. The whole trip was special, and the birthdays themselves were unforgettable. Since then, the memory of our Italian birthday celebration has been relived many times in conversation and in looking at pictures from the trip. The four of us had such a ball, in fact, that my mother recently made a point of mentioning that another set of milestone birthdays was coming up, and that she would love to make a return trip.

Consider How You Celebrate

———————◆———————

Your SO has a major birthday coming up, and your wheels are turning. You love parties, especially surprise parties. Before you start planning one for your SO, though, think carefully: On your way to someone else's surprise party, has he or she ever said, "I don't like surprise parties"? Or maybe even, "I don't want *any* party at all for my [fill in the blank] birthday"?

Your partner is just saying that, you think. Really, it will be lots of fun. Wrong. This is a time to take your SO at his or her word. When you begin planning a party for someone else, make it the kind of party *they* would like, not the kind *you* want.

Valentine's Day and Mother's/Father's Day

Three days jump off the calendar pages as days *not* to be overlooked. Forgetting them always seems to be forgiven but never quite forgotten. On the other hand, when I acknowledge Valentine's Day and Mother's Day in a sincere, thoughtful way, it brings a smile to my wife's face that lasts all day and beyond.

I admit that I've been guilty of not always making the big effort around Valentine's Day, but I also know my wife loves it when I do (and vice versa). Last year, we and three other couples were on a Florida golf trip that extended over Valentine's Day. One of the women with us—who happens to be my sister-in-law—was actually born on Valentine's Day. As a result, my brother has that holiday knocked. This time, we all decided to do something special on the evening of February 14 to celebrate both my sister-in-law's birthday and Valentine's Day. Besides being a chance to share a pleasant time together, this group celebration also ensured that Valentine's Day got the other three couples' full attention for a change. We ended up going out to dinner and then seeing a performance of Cirque du Soleil. The meal was great and the performance was eye-popping. Best of all, I discovered how much fun it is to use Valentine's Day as an excuse to plan something special and to get out and have a good time.

The same goes for Mother's Day and Father's Day. These days are perfect opportunities to make something happen for your SO and your relationship, to break out of the everyday routine and celebrate your life a little. Over the years, I've come to realize that the most important thing about celebrating any special day is attitude: Instead of looking at these days as a chore—something not to be forgotten, and then to be forgotten as quickly as possible— I'm trying to look at them as golden opportunities to have a good time with my wife, and possibly with some good friends as well.

"It's Tuesday"

One of the neatest suggestions I've heard for spicing up the otherwise humdrum routine of daily life is to *invent* a reason to celebrate. What's more, the invention doesn't have to be elaborate to be successful. One of our survey respondents recounted the following exchange: "He said, 'Let's go out to eat.' I asked, 'What's the occasion?' He said, 'It's Tuesday!' It was *so* romantic. . . ."

Examples of ad hoc anniversaries and events you and your partner can celebrate include

+ Your first date was exactly four weeks ago.

+ Your SO finally finished that big project at work.

+ You came in second in the club golf tournament— break open the champagne!

+ It's a full moon tonight.

+ It's been one year since you quit smoking. Congratulations!

7

LADIES SNORE, TOO
(BEDROOM ETIQUETTE)

Our *Bedroom . . . or* My *Bedroom?*

The bedroom is a special place. It is our private refuge, both as individuals and as a couple. Our survey respondents describe their bedrooms as their haven, their safe zone, an inner sanctum where they can leave their problems at the door. For couples, it is also the place where they face each other alone, and are most intimate—a place where they share themselves with their SO in uniquely personal ways.

For starters, we sleep here. And when we sleep, we're incredibly vulnerable to exposing ourselves in potentially embarrassing, if not humorous ways: We snore, we talk in our sleep, we contort ourselves into ridiculous, not to mention revealing, positions. What's more, we do these things with the full awareness that our bed partner can see or hear us. Trusting that other person to respect us when we're at our most vulnerable is a mark of our willingness to share our lives with someone.

So in the bedroom, we share—ourselves and our space. Sometimes, though, this sharing collides with differing needs:

"I'm awake; he's asleep. I really want to read, but if I turn on the lamp I'm going to wake him up."

"She's like a furnace, and I'm freezing."

"Why does he think it's okay to leave his dirty clothes all over my bedroom floor?"

When these kinds of conflicts arise, all too often the room ceases being "our" bedroom and turns into "my" bedroom. And that's when trouble begins. Once you begin thinking in this way, your special refuge is in danger of becoming a source of bickering instead. That's why maintaining harmony in this room, above all others, should be a top priority. Toward that end, here are a few bedroom hot spots that every couple will encounter at some point, and some tips on navigating around them.

Snoring

Have you ever woken up in the middle of the night to the sound of a train rumbling through your bedroom, only to realize that what you're hearing isn't a train—it's your SO? Your first temptation will usually be to derail that locomotive as fast as possible. Some people give in to this impulse and don't hesitate to let fly with an elbow to the ribs or a swift kick to the legs. Fortunately, they are the exception.

When we asked our couples survey subjects how they typically react when confronted with a snoring bedmate, some re-

Hope for Sufferers of Snoring

◆

One of the most effective ways to stop the snorer is to convince him or her to take steps to curtail the snoring. Nasal and oral anti-snoring sprays are sold in most pharmacies, as are Breathe Right strips—those white adhesive bands that pro football players often wear across the bridge of their nose during games. These strips help expand the breathing passages in your nose, and many couples have found that sticking one across the nose of the partner in question just before retiring will stop (or at least mute) any snoring. Even more effective is a custom-fitted mouth guard that repositions the jaw and stops the snoring. If you are the snorer in your relationship and it's causing problems, consult your dentist and find out if one of these devices can bring harmony to your bed. Finally, if either one of your suffers from persistent snoring, night after night, check with your doctor—you may have sleep apnea, a breathing disorder that interferes with sleep and can lead to other serious medical conditions.

ported they simply tell the snorer roll over. One out of four respondents said they give the snorer a nudge, and that seems to do the trick. Others employ a push, a gentle tap on the shoulder, a poke, or, in the case of one aggrieved person, pinching the nose shut. (Personally, I get a gentle nudge or a simple "Peter, turn over." It seems to work—and the funny thing is, a simple tap on my wife's shoulder works wonders for bringing down the decibel level of her snoring, as well.)

Then again, not everyone takes action. An amazing seventeen percent of snoring sufferers simply ignore their partners' freight train. Some manage to remain obliviously asleep through it all. Others leave the room. In fact, when it comes to snoring, one can't help thinking back to the old tradition of separate bedrooms. Perhaps our forebears had more sense than we give them credit for.

Virtually all of our respondents did agree on one thing: While they wanted to solve the snoring, they clearly realized this wasn't the sort of problem they were willing to let grow into a fight. Regardless of the specific strategy for coping with a snoring partner, the unifying theme throughout the responses was: whatever you do, do it *gently*. Remember, your partner isn't snoring because he or she wants to. The key is to solve the problem without letting it *become* a problem. Use a soft nudge instead of a cuff upside the head, along with a murmured word or two, rather than a grouchy tirade—and then work together to pursue other, more permanent solutions (see "Hope for Sufferers of Snoring," page 73).

Who Snores? Not Me!

◆

About ninety percent of the respondents to the couples survey were women. Among this group, the question "What do you do if your significant other snores?" brought many gentle remedies for dealing with their male SO's snoring. But when we followed up by asking respondents how their SO copes with *their* snoring, nearly one out of three vociferously denied that they could ever be accused of such a thing. "I *do not* snore!!" wrote one indignant partner. Right!

Stealing the Bed Covers

There's nothing I hate more than waking up in the middle of the night shivering, with no covers whatsoever, and glancing over to find that my wife's got them all—or, worse yet, that my dog, Navy, has managed to take over all the covers and is fast asleep on them. Fortunately, all it takes to reclaim the blankets from the dog is one swift shove and a muttered "Get off the bed!" I haven't tried that with my wife—and don't plan to.

Unlike the snoring issue, where a sense of compassion about

the problem pervaded the answers to our survey, compassion seems to fly out the window when it comes to stealing covers. The most commonly proposed remedy we got was "steal them back." The one thing I know is that if I do steal them back, I'd better be careful how I do it. I need to make sure she has sufficient covers left on her—otherwise, it's going to be a tug-of-war all night long.

Other respondents reported they take the bull by the horns and nudge their partner awake, or simply lie there and freeze until morning.

A quick solution that avoids bruises is to purchase larger-size covers and sheets for your bed. Somehow those extra few inches of comforter or sheet really make a difference.

The ultimate and truly considerate solution, however—for each of you, and for your relationship—is to expand the pie instead of fighting over it. Peace finally came to our bedroom the day we bought a king-size mattress and frame. I highly recommend one for any couple. Now there's plenty of room for both of us. In fact, I can sleep without worrying about bumping her. Best of all, now when the dog sneaks onto the bed, there's room for her as well.

Watching TV or Reading

Most of us, it seems, are tired enough at night that our partner's reading light or a little noise from the television isn't enough to prevent us from sleeping. Once in a while, however, a problem can jump up and catch you by surprise. This happened to me recently, and it was only afterward that I realized how badly I'd handled the situation.

Here's the scenario: The 2004 American League Championship Series. Yankees versus Red Sox. Midway through game 6. The season is on the line. My wife and I hop into bed, cuddle for a few minutes, and then she quickly falls fast asleep.

Keeping the Bedroom Clean (Listen Up, Guys)

◆

Going by our couples survey results, the bedroom seems to be a microcosm for the relationship as a whole. Despite what you see on television or read in the gossip magazines, many couples *have* figured out how to communicate and compromise in order to build a life together. The couples survey also shows that couples have learned how to share the load to a much greater degree than I would have thought.

Key case in point: The Post Men's Etiquette Survey found that one surefire way for any man to ingratiate himself with his SO is to pick up after himself in the bedroom—and in particular, to pick up and dispose of his dirty clothes after taking them off, rather than leaving them in a trail from the door to the bed. Apparently, the male significant others of our couples survey respondents have learned this lesson well: According to the survey, more than half— fifty-four percent—of these partners share equally in the task of picking up clothes and putting them in the hamper, while another twenty-eight percent are at least willing to pitch in and help out.

Sweet. Now I can watch the game.

Big mistake. Through her drowsiness, my wife hears the game, the excitement, the drama of it all. And she wakes up. This is a problem, because she can't fall asleep easily, especially when she's already been down for the count once only to have her slumber disturbed.

Midnight. The game's over—another clutch win for my Sox. "Good night, dear," I murmur. "You gonna read for a while? Can't sleep?" My misplaced concern is like gasoline on a fire. My wife nails me with a withering stare. Message received. I turn over. I slept like a baby that night—but I don't think she did.

If I had been at all conscious about what I was doing, I would have quietly gotten out of bed and gone downstairs to watch the game, leaving her to sleep peacefully. Interestingly, thirteen percent of our couples survey respondents say that's exactly what they do in a situation like this.

Bottom line: The bedroom is for being together and sleeping. If there's a conflict, the needs of the partner trying to sleep take precedence. So what can you do if your partner is out cold but you're still awake—besides heading to another room? Our respondents reported a variety of solutions for this problem, including

+ Buying wireless headphones for the TV

+ Getting eyeshades and/or earplugs for the sleeping partner

+ Banishing television from the bedroom

+ Using a book light when reading in bed

+ Most important of all, communicating—nicely

ETIQUETTE IMPERATIVE

Let 'Em Sleep

◆

When one partner is trying to sleep and the other is awake, the needs of the sleeping partner take precedence.

Sex and Etiquette

In response to the couples survey question "What do you do if your partner is interested in sex and you aren't?" one respondent wrote, "Hmmm. This is an etiquette book, isn't it? I'd rather not say."

Fair enough. And I'd rather not know intimate details. Frankly, what actually goes on behind your bedroom door is the business of you, your SO, and nobody else. This section is not about the etiquette of how to do it—whatever *it* might be. As far as physical intimacy goes, the etiquette is very simple: Whatever you do, whenever you do it, both people should be comfortable and be engaged of their own free will. At that moment when we are touching each other and are totally exposed to our partner, treating each other with the utmost consideration, respect, and honesty (benevolent honesty; see "Honesty," page 17) is the only guideline we need. This is etiquette at its most fundamental level, and it is critical to the long-term success of any relationship.

What *is* of interest for our purposes here, though, is how people treat each other leading up to the encounter and afterward: What have you done to make approaching sex interesting and exciting, and what have you done or said after the fact to reinforce the intimacy of the act? Etiquette can make a big difference in dealing with such questions as . . .

+ How do I say no when I'm not in the mood?

+ What can I do to set the stage this evening?

+ The Yankees and Red Sox are playing tonight. How long do I have to wait *afterward* before I can turn on the TV?

Whenever two people come together and interact, you have etiquette. Remember, this is how Emily defined etiquette. What you do in the moments before and after are some of the most critical interactions you will have as a couple. Another note: Our couples survey also found that most couples make a point of being very consistent in how they treat each other as they approach being intimate and afterward—indicating that compassion and thoughtfulness matter more than novelty.

The Etiquette of the Prelude

Call it anticipation. Call it setting the mood. Call it whatever you want, but know one thing: it *matters*—for women *and* for men. We all know a man can be ready at a moment's notice, but that's a physical thing. Building emotional desire is something both sexes need, and it comes in all sizes and shapes. Suggestions from couples survey respondents included:

+ Doing the dishes (partners are actually viewed as more desirable when they help with chores)

+ Kissing. I *love* to kiss.

+ Back rub is the best, sex or no sex

+ Baths (our survey's number one–ranked mood-setter— apparently lots of people enjoy a nice soak as a prelude to sex)

+ No kids around

+ A full body massage with lotion or body oil

+ Talking about it

+ Cleaning the living room

+ Being playful and affectionate with each other

+ Dinner and dancing

The list goes on. The common thread here is that *the couple is acting together* to build a mood for the moment. Building emotion together while acknowledging your feelings for each other—that's the key. One respondent's description of a memorable evening with his wife summed this concept up perfectly: "One time we watched our wedding video and it made us all

---

ETIQUETTE IMPERATIVE

Sincerity Matters in Sex, Too

◆

Where sex is concerned, communicating sincerely is the most essential element in setting the right mood.

---

emotional again, and we reenacted our wedding night all over again. That was pretty cool."

It doesn't always have to be a specific thing, like a bath or a massage, that makes the difference. "He's very complimentary in a sincere way and always makes me feel great about myself. That is the best prelude—feeling loved and beautiful," wrote one woman. She got it right. The bottom line in setting the mood is: Communicate and be sincere about it. Remember, you can talk the talk, but if it doesn't come from your heart, your SO will know it. And then you might not be walking the walk.

Sometimes the Mood Just Isn't There

There are no sure things in this world. Just because you help with the laundry or cuddle up to him throughout the day or fix a beautiful meal or light candles and set a romantic mood, sex isn't guaranteed that night.

We've all navigated the murky waters of "I'm interested"/"Well, I'm not." The key is to do it gently, compassionately, and honestly.

If you're the "I'm not" party, don't tell your SO to go take a cold shower. Instead, take an approach that demonstrates your understanding. "Explain why," wrote one respondent, who asks "if we can wait till a time soon after. We always respect each other's decisions."

That's the heart of the matter. Respect. On both people's parts.

Another option cited repeatedly in the couples survey: "Ask for a rain check." With a little thought, you can even turn the rain check into a quite enjoyable event: "Make a date with him and then pull out all the stops."

What's fascinating is how many couples survey respondents stressed the idea that getting down to having sex is a mutual decision. People learn to read each other and begin to know which actions and moods are indicative of interest and which aren't. "We are pretty tuned in to each other," wrote one respondent, "so we can tell when the other might not be receptive to advances and just don't push it."

Don't assume, by the way, that men are always the ones instigating an encounter and getting turned down. Male respondents acknowledged that they can be the person saying no, while a number of women allowed that they're sometimes the party doing the instigating.

Many couples say that compassion and compromise come into play here as well. Forty-four percent of our respondents indicated that at least sometimes they will make the effort anyway, even if they're not really interested. The bottom line: No single answer is right all the time. Mood, attitude, circumstances may all combine to make a "No, not tonight" the right choice one time, while a willingness to go along with the more amorous partner may be the best answer another time. Underlying the choice is an effort to be considerate and respectful not only of your SO's feelings but of your own feelings and situation as well.

ETIQUETTE IMPERATIVE

You Matter, Too

◆

Relationship etiquette isn't only about being considerate of the other person. It is about being considerate of yourself as well.

The Happy Bed

On the theory that a couple that sleeps well is a happy couple—or at least a more cheerful couple—take stock of your bedroom. Does it provide the amenities that make it a place conducive to sleeping and other activities? The king-size bed was our biggest improvement in the atmosphere for getting a sound night's sleep. But funny enough, a fresh coat of paint last fall also made an amazing difference. The room seemed brighter and more friendly after the paint dried and a new set of curtains went up.

Here are some other suggestions for quickly and easily upgrading the mood in your bedroom:

+ Clean, soft sheets with a high thread count
+ Comfortable mattress and a couple of extra pillows
+ Sound machine—good music or nature sounds can really make a difference
+ Quality shades/drapes
+ Not having a clock visible from your bed (which encourages clock-watching at night)
+ Night tables or other surfaces where a couple of candles can be strategically and safely positioned

Afterward

You and your SO have just shared the ultimate intimate moment. You mutter, "Thanks," and turn your back on your partner. Kinda takes the wind out of the sails as far as the big event is concerned, doesn't it?

What to do instead? Simple: Pay some attention to your SO afterward, just as you do before and during your encounter. A little thought before you act will go a long way toward not only ending the event on a high note but also encouraging your partner to be interested and eager next time. Again, the key is to be conscious of how your partner is feeling and of how your actions will affect your partner. Here are a few things you can try instead of turning over to sleep, turning on the TV, lighting up right away, or heading into the kitchen in search of a sandwich:

+ Snuggling

+ Cuddling

+ Holding her

+ Kissing him

+ Talking to her

+ Sincerely thanking him for a wonderful time

ETIQUETTE IMPERATIVE

Get Clean

The considerate bed partner brushes and flosses before retiring, doesn't wear dirty socks or other smelly clothes to bed, and takes a pre-bedtime shower if necessary.

8

CHILDREN CHANGE
EVERYTHING

ON JUNE 12, 1979, MY LIFE CHANGED FOREVER, AND SO DID my wife's. That was the day our first child, Anna, was born. When I say that her birth changed my life, I'm not just talking about lost sleep and dirty diapers. Up until then, I'd held a number of different jobs, from being a reporter on a local newspaper to teaching art to high school students. When Anna arrived, I suddenly realized I had to stop testing the waters and commit to something that would bring in a steady income. Fortunately, I found a perfect job at a small college and, just like that, settled down to being a responsible parent and spouse.

Relationships and Pregnancy

Of course, pregnancy is a major event for a woman, but it's also a major event for her partner—and for their relationship with each other as well. Each party has to deal with the physical changes that are happening to her, and how those changes are affecting both of them. They also have to begin preparing to deal with the impact that an additional small person is going to have on their lives and how they interact with each other. If ever there was a time for communication, compromise, and commitment, this is it.

For Her

For a woman, the physical and emotional changes of pregnancy are very real. You want to still be yourself, but that self is in flux. You may feel ill, especially in the earlier stages of pregnancy. Your stamina will be drained, your patience tested, your feet and back often aching.

The result is, inevitably, stress—not only from the changes that are occurring but also from the growing realization that your life will be irrevocably altered once the baby arrives. These stresses are felt especially keenly with a first pregnancy, but *every* new addition to the family has its own unique, life-changing impact.

Now, stress, as we know, is a perfect fuel for short-tempered, harsh behavior, even when a person doesn't really mean to behave in this way. Nobody is going to deny the inherent difficulties of going through a pregnancy and how this can affect the way the woman feels. From the expectant mother's standpoint, the key to sailing through these waters with as little upset as possible is to be aware of how the changes may affect the way you deal with the people around you.

For Him

For the expectant father, patience and understanding born of your commitment to your partner are the keys to navigating the pregnancy. She's going to need you to be understanding and supportive, and to pitch in and help in ways you may never have done before. If you haven't already, this is the time to . . .

✦ Perfect your vacuuming and housecleaning skills

✦ Polish your ability to cook at least a rudimentary meal or two

✦ Learn how to sort, wash, and dry the laundry

✦ Pin down the finer points of shopping for groceries and other household items

Your interest in what is happening to your wife is also important. I can recall looking at books together, learning what stage of growth we were at with each passing week, and marveling at the first sensation of feeling the baby's movements.

You can't carry the baby, that's for sure. But you *can* share the experience and be a partner in it. And that's essential for both of you.

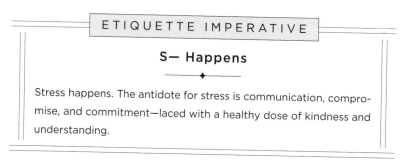

ETIQUETTE IMPERATIVE

S— Happens

Stress happens. The antidote for stress is communication, compromise, and commitment—laced with a healthy dose of kindness and understanding.

For the Two of You
This is a time to begin refocusing your priorities and to truly learn the meaning and importance of compromise to your relationship. With change comes stress, as new roles are defined and as new priorities emerge. And the antidote for stress is communication and understanding.

As you communicate about what is happening to each of you, as well as to both of you as a couple, you'll find yourself getting better and better at anticipating upcoming challenges and planning how to deal with them. By being considerate of each other's changing situation, you can head off the stress that comes from misunderstandings or from being poorly prepared for what lies ahead.

"There's No Time for 'Us' Now"

Yes, you have kids, and inevitably they become the priority in your family. Still, when I listen to parents talk about the effect their children have had on them as a couple, I'm struck by how quickly the kids become an excuse for the parents to stop taking time to focus on each other.

The reasons for no longer spending time with each other are many: sitters are expensive; every ounce of energy is focused on getting through each day; it's more important to spend quality time with the kids; there simply isn't enough time for "us" now.

One of the biggest rationales for putting the kids before your relationship with your SO is that there will always be time for "us" later. Sharon and Bill's approach is typical: "One compromise you have to make with children is that you might not be there for each other now, but later on you'll be there for each other," they told us. "When the time comes, we know we'll be there eventually. The kids have to come first now."

The big mistake Sharon and Bill are making here is in thinking that raising children and focusing on themselves is an either-or situation. There's a world of difference between kids coming first and kids dominating your life to the point where there's no time for the two of you. Successful couples work hard on bringing up their children, yes—but they also work hard on their *own* relationship at the same time. Not only is the effort beneficial for the parents' relationship, but by modeling this positive behavior, they're starting their children out on the road to successful relationships, too.

Take Tim and Andrea: "We used to go out on the town every Friday night, but not since the kids came along," they reported. "Now, we might go for a walk instead." Your family will only be as strong and stable as your own relationship. This is the time, when you are in the throes of raising children, that it becomes more important than ever to carve out time for just the two of you.

Tim and Andrea have the right idea: Maybe they can't go out for a big night on the town the way they used to when it was just the two of them, but they can still find ways to steal time for themselves.

On my way home from work, I often see one local couple out on their daily walk together. They're always out there at the same time each day, their dog alongside them, walking at a good pace, clearly getting in some terrific exercise—and they're always talking.

Date Night

One of the best things couples with children can do for their relationship (and their sanity) is to have a regular date night that they plan and hold sacred. That means, first of all, finding a reliable sitter who can commit to watching your kids one night every week, or every two weeks, or every month—whatever pattern works for you.

The Regular Sitter:
Your Relationship's Most Valuable Ally

◆

When you're raising young children, a regular babysitter is one of the best allies your relationship will ever have. Angela began to sit for us when she was just 12 years old. She lived just down the road, and she would come over to help out on an afternoon or (as she got a bit older) to tend to the girls when we went out for an early evening or if we were visiting friends who lived close by.

In the neighborhood, Angela became known as "our" sitter. She had a regular job with us, and we had the peace of mind that came from knowing we could count on her to be available. We even took Angela on our vacations with us. Those two-week vacations at a family summer home were a dream come true. Thanks to her, the kids were completely taken care of—played with, fed, put to bed. As a result, we were free to spend some serious quality time, both alone with each other, and with the kids and Angela.

Later on, when our kids no longer needed a sitter, Angela would call at holiday time and invite them to spend the night of December 23 with her—a girls' slumber party and a chance to catch up. She's married and has children of her own now, but she'll always have a special place in our family, too.

If you can't afford a sitter, try asking your own or your SO's mother or father to step up to the plate, or work out a swap with some friends who also have young children—you sit for their kids once a month, and they do the same for you.

When you do go out, make it a point to get away to someplace where you can really focus on each other, a place where you can talk and renew your infatuation with each other.

Harder to pull off, but perhaps even more rewarding, is for the two of you to take a vacation on your own. When our girls were about 7 and 4, our closest friends made us a proposition.

He had received some free tickets to fly anywhere in the United States, and he suggested that we all fly to San Francisco for a long weekend in the wine-growing Napa Valley region. We went for it: We arranged for the kids to be taken care of for the four nights we'd be away (I've talked to numerous couples who have done this, by the way, and I was interested to note how, once they commit to going somewhere on their own, they *always* managed to find someone—a parent, friend, brother, sister, grandparent, or sitter—who could take the kids) and flew west.

The trip was sensational. We stayed at a small inn in the Napa Valley one night and at a hotel in nearby Sausalito for three nights. We got to visit with my wife's brother, who lives out there. We ate great food, including an incredible lunch at Napa's Domaine Chandon (we still talk about the red raspberries and sauce, served in a perfect small bag made of luscious dark chocolate). And we came home renewed.

That was when we realized that carving out some time just for us made us better parents and was valuable for our children as well.

ETIQUETTE IMPERATIVE

"There's No Time for Us!"

◆

I cringe when I hear couples say they don't have time for each other. Children should never be used as an excuse for not dealing with each other. A successful couple works just as hard on their *own* relationship as they do in raising their children. When relationship issues go unresolved, they fester and grow—until finally they can't be ignored. And when that happens, your problems will take far more effort to resolve than if you'd dealt with them initially.

Resolving Child Care Inequity

Invariably, it seems, one partner tends to take the lead in child care. The most valuable thing the partner who is *not* the primary caregiver can do is to jump in and help out whenever he or she is at home. Rather than retreating to the newspaper or TV, a thoughtful partner will take the little tykes off his or her SO's hands for a morning or an afternoon or a whole day on the weekend, and will also make sure his or her SO has time to spend on a favorite activity or simply to catch up on some much-needed rest. The hours you put in at work will only be resented if you don't make a consistent effort to carry your weight when you're home.

Many of our couples survey respondents emphasized how important it is for both partners to pitch in:

> *"I'm a stay-at-home parent. It's my job, and I'm lucky to do it. However, I do need relief from time to time, and he's happy to do that."*

> *"He helps as often as he can by bathing/feeding/ dressing/putting our toddler to bed."*

> *"When we were first married and the baby came along, I was working more hours than he was, and so a lot more of the child care fell to him. My mother was astonished that not only could he change a diaper as easily as I could but he also was a whiz at folding diapers."*

> *"When our children were babies, I worked the weekend program at the hospital and he was the 'mom' the entire weekend. I worked 12-hour shifts when we had a three-year-old, a one-year-old, and an infant, so he 'bonded' with our children! When he started his own business and was gone for several days at a time, I took over. . . . He has a great relationship with our children. I am never nervous about leaving them with him for a weekend."*

ETIQUETTE IMPERATIVE

Being Considerate to the Nth Degree

◆

Whatever the setup, a family is successful when child care is viewed as a shared responsibility, and when each partner believes that the other is contributing to the best of his or her ability under the circumstances.

In my own case, when I started working for a small college shortly after we had our first child, I knew I couldn't be with my daughter all day, but when I did get home I wanted to be with her and take care of her, instead of just vegging out on the couch. My efforts met with mixed success: More than once on a Saturday afternoon, my wife would find me asleep on the couch, a golf tournament on the television and my daughter sound asleep on my chest.

I believe the vast majority of parents in my situation actually relish the time they're able to share with their children, and come home ready to help even if they're wiped out from work that day. One of my most vivid memories is of a Saturday when my oldest daughter was about six months old. I'd already agreed to take care of her so my wife could spend the day with some friends. As it turned out, however, I had to go in to work to finish a project. Rather than abdicate my role as caregiver that day (and ruin my wife's day in the process), I took Anna with me to my office. I put her car seat on my desk, then sat down and managed to get everything done I needed to. Work was quiet that day, and Anna fortunately was an angel.

I had just about finished up when the president of the college happened to walk by. He saw us together and came in, and we chatted for a while about families and kids. Not only did I get my work done *and* manage to take care of Anna at the same time but I also ended up having some prime face time with the boss.

Get Clear About Responsibilities—Ahead of Time

Your SO is not a mind reader, and neither are you. If either of you is unclear about who is responsible for what, you need to communicate about it. And the best time to communicate is when you can have a sane, rational conversation—not while pots are boiling over in the kitchen at the same time you're trying to give little Suzie her bath. Yelling at that moment might resolve the immediate situation, but frankly, I'd much rather avoid being in the situation to start with. To do that, you both have to have a clear understanding of your roles and needs. The only way that will happen is to sit down together and talk your responsibilities over *ahead of time*, when you both can listen and hear what the other is saying, rather than when a problem is unfolding before your eyes or when anger is welling up inside one or both partners.

Just remember, chaos does happen. By defining your roles and responsibilities ahead of time, you and your partner can stay focused on the issue at hand when trouble strikes. Afterward, when you're both calmer, make an effort to do a postmortem. Pick a peaceful, quiet moment when you're both relaxed and concentrating and you can really communicate, compromise, and then commit to each other. Talk the episode through, seeing where you went wrong and what you did right. Then set up a plan for doing it better the next time. And don't forget to start out with a sincere apology when it's warranted.

That's etiquette. That's problem-solving (see "Four Steps to Resolving Disagreements," pages 146–147). That's building a successful relationship.

Discipline: The Importance of Being on the Same Page

I was standing talking with my neighbor Dennis recently when the school bus pulled up and disgorged a bunch of junior high school kids. One of them was Dennis's son.

Son: "Hey, Dad."

Dennis: "Hey, son. You'd better get inside and get to work on your reading homework. You know what your mom said this morning."

Dennis's son dutifully went inside, and we resumed our conversation. Five minutes later, his son reappeared, coat on. "All done," he said.

"What?" said Dennis.

"It's all done."

"Get back in there and do your reading. You know what your mom said—and I'm going to make sure you do it."

"But I *did* my reading." They went back and forth like this for another minute before his son broke: "Dad, I can do some tomorrow, and some more the next day," he pleaded. "It's not due until the end of the week."

"In . . . *now*. Go!" And his son slunk back into the house.

Dennis never relented. No matter how much his son bargained and cajoled, he stood firm and didn't change the rules that had been set. And his son learned yet again that Mom and Dad were a team. One wasn't going to overrule the other.

To make your discipline as successful as possible . . .

✦ **Support your partner.** If children spy a crack in the parental wall, they'll work it relentlessly. If you disagree with your partner, do so privately. Discuss your perceptions together, and then come to a joint conclusion about how the situation should be handled in the future.

✦ **Be consistent.** Once a rule has been established, stick to it. Making Johnny do all his homework one day and then letting him watch TV as soon as he comes home the next will serve only to confuse him. Kids will never say so, but they actually crave consistency from you and your SO and will respond to it in spades. When the head of one of the nation's leading child care centers was asked why she never had any discipline problems caring for 300 toddlers every day, she summed it up simply: "We're *incredibly* consistent in enforcing our rules." Because the children knew exactly what types of behavior were frowned on, she explained, they never crossed that line.

✦ **Make consequences for misbehavior clear and reasonable.** "You're grounded for life!" may sound impressive, but it isn't. The punishment is unclear and unenforceable. Therefore, it's meaningless, and you both know it. "You are grounded for two weeks." Now, *that's* a clear and reasonable consequence for a teenager who just arrived home three hours after curfew, after failing to call in as agreed.

✦ **Don't make your SO the bad guy.** "Wait till your father gets home. Then you're going to be in real trouble." Wrong.

The Empty Nest Syndrome?

As our daughters finally went off to college, I can remember friends warning us about the "empty nest effect" on a marriage. They'd use dark, foreboding terms: "You're going to be depressed" . . . "It's a tough time" . . . "You'll be making a major adjustment," and so on.

Sorry, but au contraire! My wife and I love our daughters. I wouldn't trade the experience of bringing them up for anything. I liked being a parent. I liked the responsibility, and all the together time it took (and still takes). But I've got to say, I also really like the time my wife and I now have to focus on enjoying our home and each other. I take pride in the fact that both of our daughters are now living in their own places, and doing so successfully. We haven't stopped being parents. They call all the time, and we're still involved in their lives—their ups and their downs, their fears and their victories. But they are embarked on their own lives, and that's as it should be.

After all, isn't the goal of parenthood to prepare children for being independent? When it came time for ours to test their wings, we enjoyed watching the process and being there to encourage and help. When they were young, we were responsible all the time for them. Today, we have a new and different role. By being successful parents while they were with us—by teaching them to trust us, respect us, and be honest with us, and by doing the same with them—we have been able to maintain a close relationship with both of them, even though they're no longer physically here in the house. And that, frankly, makes the separation not only bearable but also a joy.

9

FINANCIAL ISSUES

"MONEY IS THE ROOT OF ALL EVIL" IS ACTUALLY A BASTARDIZED version of a biblical passage, but it carries a germ of truth where couples are concerned. Like it or not, money is the cause of many of the most severe disagreements in relationships. It's not just lack of money that causes stress; what we choose to spend money on when we *do* have it can be a sore point as well.

Couples survey respondents saw this issue clearly:

"Her spending priorities are not the same as mine."

"He spends superfluously."

"I tend to overspend at Christmas, while my husband wants more money allowed for vacations."

"Sometimes he spends a lot of money on stupidity."

"Sometimes I spend too much money."

"Wife is a professional shopaholic."

What seems to be missing in all of these examples is the one cure for financial disagreements: a combination of *communication* and *compromise.* Both of you have to know what your resources and your expenses are. In addition, you have to work together as a couple to develop a process for meeting your responsibilities that's clear and fair to both of you. If any of these three points are not fully understood by both partners, then trouble is brewing.

Bank Accounts: Joint or Separate?

When we first got married, my wife and I had separate checking accounts. Thirty-two years later, we *still* have separate accounts. Frankly, I can't imagine trying to manage a shared checking account. As it is, I have a hard enough time just making sure all of my entries get written down in my own checkbook. If I forget to jot a payment down and it screws me up—well, tough on me. I have to deal. But with a joint account, failing to record a payment (or even simply failing to apprise her of a payment I *did* record) could cause my wife's check to bounce, and vice versa. Since neither of us wants to have to keep track of the other's expenditures, his-and-her checking accounts make perfect sense.

Given our own arrangement, I figured that very few people, when asked by the couples survey whether they have a joint checking account, would answer in the affirmative. Was I ever wrong:

Type of Account	Percentage Answered
Joint checking account	43.9
Individual checking accounts	25.9
Joint and individual checking accounts	23.4
None of the above	6.9

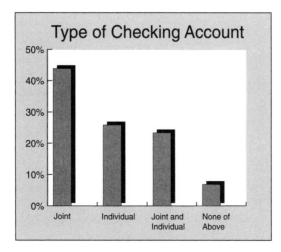

As the accompanying graph shows, sixty-seven percent of our respondents—two of every three—have some kind of joint checking account, and almost half (forty-four percent) have no individual account in their own name whatsoever. At the same time, no more than a small handful of respondents reported any problems in the relationship due to a joint account. Clearly, couples are making joint accounts work, and my hat is off to them.

That doesn't change my view on the matter, though: My advice is to maintain separate accounts, and to clearly delineate who pays which bills (see "Flashpoint: Who Pays for What?" on page 102). If you and your SO want to keep an eye on what each of you is spending, simply review your bill payments together each month.

The one potential problem with separate accounts is if one partner feels the other is hiding something. (Of course if that's the case, the trust that's missing from the relationship is a much bigger deal than the issue of the checking account.) Our use of separate accounts has nothing to do with trying to hide our income or our spending from each other—it's simply our way of avoiding screwing up each other's accounting.

FLASHPOINT

Who Pays for What?

◆

If you earn twice as much as your SO, should you contribute twice as much to your housing costs, telephone, groceries, and other essentials? Will the cable bill be in your name or your partner's? And who pays when a married, two-income couple goes out to dinner?

The bottom line is, *it doesn't really matter*, as long as your budget plan is based on communication and compromise. In fact, our couples survey found that couples were able to make a wide variety of arrangements work, provided they were both in agreement on what these arrangements should be. If, on the other hand, one person is left out of the financial decision-making process, feels that he or she is paying an unfairly large portion of the monthly bills, or senses that he or she is always paying the dinner or entertaining tab—then you're headed for trouble.

The other big problem, of course, is when you both agree on your monthly outlays, yet you still can't seem to come out on the right side of the income-to-expense ratio. This issue needs to be addressed ASAP, because mounting debt is one of the most lethal relationship stresses of all. If you find that you're falling behind financially, consider contacting a financial planner and/or credit consultant in the near future.

Stress and Finances

When the financial going gets rough, couples can easily become trapped in a vicious circle in which stress leads to rudeness and blame and rudeness and blame lead to stress. The best antidote for this vicious circle is to administer a good dose of etiquette. Why? Because etiquette encourages us to *communicate in a way that resolves situations* and *builds relationships* instead of tearing them down.

When it comes to money issues, one key step is to identify

exactly what your sources of financial stress are. Uncertainty and lack of action only add to whatever stress you may be feeling. Once you and your SO have determined where your financial stress is coming from, you can take proactive measures—and avoid getting sucked into the stress-rudeness-blame vortex.

When asked to name the main source of their financial worries, our survey respondents overwhelmingly cited one of two things:

+ Not having enough money

+ Financial planning for the future

When you deal with these issues head-on as a couple, using forethought and communication, stress on the relationship is reduced. The key is to focus all communication on solving the problem, rather than on attacking each other. Say, for example, you're discussing the fact that you and your SO are barely scraping by from paycheck to paycheck:

+ **Wrong:** "You need to get a better-paying job. We can't survive like this." Accusing the other person of being the source of the problem without offering a solution will get you nowhere. Instead, stay focused on the problem itself.

+ **Right:** "I've been looking at our spending and our income. We need to work together to cut our expenses so that we can live on what the two of us are making. If that isn't enough, we may have to think about ways to increase our income. I've categorized our monthly expenses. Let's look at them together." This approach keeps the focus on the problem and offers a solution that both of you can contribute to, rather than defining the issue as just one person's problem.

In the End, It All Balances Out

◆

Gifts are about giving, not about value—so get out of the value gutter. As for my wife and me, some years she's been extravagant and I haven't. Other years, it went the other way. No matter which side of the equation each of us happened to be on, it was the appreciation of the gift that mattered.

A couple of years ago, there was a box with my name on it under the Christmas tree. I hadn't asked for anything in a big box, and I couldn't imagine what it could be. The present turned out to be the most ridiculously outrageous coffee/espresso maker you ever saw—something I wouldn't have bought in a million years. Yet there it was. Kiss, kiss. "Thanks, honey! This is awesome." I set it up right away. I now use it every day to make her coffee, which, as you know, I bring to her in bed every morning, and guests love the real espresso that the machine makes perfectly every time.

I know I didn't get anything near the value of that espresso-maker for her that year. But it wasn't about value; it was about pleasure. And we both know that sometimes the gift-giving shoe is on the other foot, and in the end, it all balances out.

Making Decisions the Hard Way

In the short term, the easiest way to make a decision is just to make it, regardless of what anyone else thinks. I could decide to buy a pickup truck today—go to the showroom, sign the papers, plunk down some money, and commit my wife and me to four or five years of payments. But it's pretty clear that I would also be committing a serious relationship error.

The alternative takes more work, but the effort is well worth it: Talk with your SO about finances—what you'd like to buy, what you can afford, and what your goals are, both now and in the future. The two of you have come from different places and you're each used to doing things a certain way. Blending your different approaches to finances and money takes time and requires ongoing communication and compromise.

So if I really want that pickup truck, I'll need to talk to my wife first. I always approach this sort of "big purchase" discussion the same way I go about making a proposal at work: I define the current problems and needs, and I propose a solution. Next, I explain how my proposed solution (in this case a new truck) can be implemented without busting our budget. Finally, I ask her to look my proposal over and think about it, and we agree to talk about it again later.

If the big purchase is something my wife feels we need, then our roles are reversed and she makes her pitch to me. This give-and-take approach accomplishes three crucial things:

1. Each of us shows respect for the other by treating the other person as an equal partner in our life together.

2. We've learned through this process to respect each other's opinions and way of looking at our problems and needs. My wife is often able to refine and clarify the issues I bring to the table, and I like to think that I do the same.

3. The process works equally well for both of us. That's a real partnership.

In the Post Couples Survey, a clear majority of respondents felt that all decisions related to money should be shared. And when it comes to major decisions such as buying a home, seventy-seven percent—more than three-quarters—of our respondents said that these decisions should be made by both partners. (Interestingly, those people who said they deferred to their SO, or vice versa, also indicated they were comfortable with that nonsharing approach and had no complaints about it.)

In reviewing the Post survey results, I was struck by just how many of our respondents report having strong, positive relationships. To me, it's no coincidence that a large majority also recognize how vitally important it is to confer about major decisions, especially financial ones. The importance of sharing in major financial decisions could not be clearer—and the only way these decisions can be shared is if you and your SO are spending time communicating with each other.

ETIQUETTE IMPERATIVE

Talk

"He's so uncommunicative." . . . "He doesn't like to talk."

One recurring negative complaint heard in the couples survey had to do with husbands or boyfriends who are unwilling to sit down and discuss key issues, particularly money matters.

Talk, especially about finances: what you're spending, what you're making, and how you're going to make ends meet. You both have a huge stake in your financial health. Respect that.

Talk to avoid the arguments that are certain to happen if you don't.

Talk now to nip any financial problems in the bud while they're still manageable, rather than waiting until they overwhelm you.

10

Divvying Up the Chores

AS FAR AS I'M CONCERNED, CHORES STINK. BUT THEY ALSO need to get done. And just because my wife does some task in my place doesn't make it stink any less. So, while part of me would love to sit back and let my wife "magically" do everything around the house and yard, the considerate part of me realizes that this is unfair—and also that eventually, her reaction to this situation will be, "Why am I married to this guy?"

This is where reality intrudes: When it comes to our household chores, I've got to contribute and she's got to contribute. That means deciding who's going to do which chores. Otherwise, my wife's going to end up doing them all, and then she's going to start asking *that* question.

Your choices come down to this: You and/or your SO can do the chores; you can hire someone else to do them; or you can try

Appreciating the Effort

———◆———

From one of our Post Couples Survey respondents: "Basically, he does well over half, but not necessarily very well. I often *redo* a good portion of it, without complaint, glad that he has helped. Includes crusty dishes (no dishwasher), balled-up clothes that now need ironing, refolding towels (he'll never get that), putting away things my way, etc."

She's amazing. She appreciates the effort.

putting your kids to work. Some chores—such as housecleaning, yard work, pool cleaning, child care—can be hired out (and often are), but you still have to manage the hired staff, in addition to doing all the other chores that you aren't paying to have done.

Children are a great resource, and they should be expected to help out around the house, but take a reality pill before you go too far down this road. There are limits to the chores that a ten-year-old—even a motivated ten-year-old—can accomplish.

In the end, all chore scenarios lead to the same outcome: You and your SO are going to do the vast majority of them. So get ready to deal.

Chores Strengthen Your Relationship

Chores make relationships stronger.

Yeah, right.

No, really.

It's kind of like making a silk purse out of a sow's ear. Instead of thinking about the individual chores you have to do and what a pain in the neck they are, think about your duties in this light: Sharing the load together, with each person taking part of the burden, is one of the best ways to strengthen your relationship.

That's what it really means to be a couple—each of you pitching in and doing things that will benefit both of you. The fact is, doing things together is a significant part of why you became a couple.

A relationship involves sharing your lives, and chores happen to be one of those things you end up sharing. If you tackle tasks collaboratively, without grousing or complaining, it makes perfect sense that your relationship will benefit. It's the complaints—the whining, the pouting, the sulking, and the outright avoidance of chores—that cause all the relationship difficulties. Think about it: Wouldn't *you* rather collaborate with someone who is consistently positive and helpful?

Helping out with the chores isn't just about spending quality collaborative time with your SO, of course. It can also be a way of showing your consideration for how he is feeling, or recognizing that she's pressed for time and could use a helping hand.

Many couples survey respondents commented on various ways their SO helped with the chores, and how much it meant to them:

> *"Cleaning up the house when he knows I'm tired."*

> *"Irons my shirts while I'm getting ready for work (not typically a boyfriend-like thing to do!)."*

> *"Picks up after himself without me asking, because he knows I like a clean house."*

> *"Since I work longer hours, he always makes an effort to do more around the house."*

Working Together

Some couples literally share the chores by pitching in and working together. My wife and I do that quite a lot. For instance, this past fall I went outside one day, determined to clean the garage. This is an especially dusty, nasty, unpleasant chore. To make matters worse, a pile of wood and debris, left over from a renovation project, sat in the middle of the garage floor, preventing my wife from being able to pull her car into the garage.

I had just begun carting the wood away and sweeping up the debris, when my wife suddenly appeared in jeans and a sweatshirt. She didn't say a word—she just pitched in and started helping out. We got the job done lickety-split, and this gave me the opportunity to help her out inside the house next.

I really liked our sharing of the chores that day. Sure, we could have each done our own set of chores, but doing them together gave us a chance to be near each other and even talk some as we worked, rather than toil in isolation.

Sharing the chores entails a combination of planning and evolution. Two respondents from the couples survey made this point perfectly:

> "Despite my complaints, I know we've really got a good balance. I have never once bothered myself with car or yard issues. He's never once had to worry that the children have the proper school supplies or clothing, or that items might not make it from Santa's list to under the tree on Christmas morning. There are scores of 'hidden jobs' in a large and busy family, and somehow we've managed to get most of the important ones done without killing each other."

> "Because the laundry and bathroom are entirely my responsibility, my SO does most of the dishes (we don't have a dishwasher) and all of the vacuuming. These are the two things I hate doing the most. All in all, I do lots more than my SO, but this is mostly my choice. I enjoy a clean house, while my SO wouldn't mind a bit of clutter. We're both satisfied in our current division of labor—but it did take us several years to reach this point."

That's What Makes Horse Races

◆

One of my father's favorite sayings—muttered when someone dis-agrees with him—is "That's what makes horse races." After reading all the disparate comments from our couples survey respondents, I've become a real believer in this maxim. Here's one example of why: Commenting on the division of labor in his relationship, one man told us, "Early on, we decided that she'd do the housecleaning and I'd take out the trash." I'd like to hear what his wife has to say about that division of labor!

To be fair, in some homes taking out the trash *is* a major chore: "He *always* takes out the trash, unless he's out of town, which is a big deal because we live in a fourth-floor walk-up," wrote one grateful SO.

How Couples Divide and Conquer

The division of chores is not always going to be a perfect fifty-fifty split. For one thing, it's hard to quantify the time and effort in-volved in each task. Can you equate paying the bills with doing the dishes? Is mowing the lawn more or less effort than raking the leaves or doing the laundry? The fact is, every situation is differ-ent. In my house, for example, mowing the lawn is a major 1.5-acre effort, and we just let the wind blow the leaves away. As the Post Couples Survey confirmed, there really is no single "right" way to split up your household chores. Instead, each couple must de-velop their own unique approach:

✦ She does the inside; he does the outside.

✦ She works two jobs; he does most of the chores and she pitches in when she can.

✦ He's the breadwinner; she maintains the home.

✦ They both work; they both pitch in.

✦ She can't stand vacuuming; he hates to do dishes.

ETIQUETTE IMPERATIVE

Don't Sulk—Solve the Problem

◆

If you feel you're getting the short end of the stick chorewise and you resent it, don't let your feelings fester. Talk to your SO about the problem. Don't accuse, and don't berate. Instead, identify the issues. Give credit where credit is due. Offer a realistic solution. You can even get creative, as one survey respondent did: "My solution for getting him to mow the lawn: affixing a 99-cent cup holder for his beer to the riding mower. Now he can combine his after-work cocktail with something productive!"

I wonder if my wife has read this yet.

Clearly, couples have evolved a variety of ways to delegate specific tasks. The real key to a successful division of labor is that each partner is comfortable with the division of chores, and with the relative amount of work each person's chores represents. Our survey results indicated that some people are more comfortable when they perceive the split to be more or less equal:

"Despite a very traditional division of household labor, I think it is equally split. We do the chores that we are each good at—he had investments prior to the marriage, so he is a good manager of household finances, whereas I do the laundry and cooking (as I have learned to do those well). . . . If I didn't cook, then we would be having pizza and macaroni and cheese every night."

"I'd say we're probably about sixty–forty right now (the sixty is me). For the most part, this issue has equalized over time. When we were first living together, I used to do everything (except his laundry), and it would drive me nuts. But now I'm not afraid to ask for help, and we work together to get things done."

If either partner begins to feel that he or she is carrying all or most of the load, trouble won't be far behind. As one respondent wrote: "I wish he would see things that need to be done. It would be nice not to have to make him lists or ask for help."

The survey also found that others are perfectly comfortable with an unequal split in which they do the bulk of the chores. Usually, this is because she or he wants things done a certain way and is willing to take on the extra work to make sure they get done properly. If this is the case, the best thing the partner can do is to step out of the way *and respect the effort* his or her SO puts in.

> *"I am a clean freak and I prefer to do all the cleaning. His job is to not make it worse, and to be respectful of it."*

> *"My husband has asked me for a year now to come up with a chores list for the house that we can then divide up, because I do too much and he wants to help. I have trouble letting this part of my life go. Although I am not a neat freak or even close, I do not like others messing around with my house or cleaning stuff. Crazy, I know. . . ."*

> *"I realized early on that I like things clean, while he couldn't care less if it was messy. Since it's important to me, I do it. When he offers to help, which is frequently, I give him simple tasks to do that he can't screw up, and I praise him for it. We agreed when we moved in that I would maintain the house and cook, and he would pay all the household expenses. I find this to be fair, and if I start to feel put upon, I give myself the night off and take us out to dinner, or take a day off from my self-imposed cleaning schedule. When I was a child, my mother had us shaking in our boots because the house didn't look a certain way, and it killed two marriages. I decided early on to keep up my standards, but to relax them somewhat and to recognize that my way is not the only way.*

Furthermore, if it has to be my way, then I should do it myself and shut up about it already."

Appreciating Each Other's Contributions

It seems to me that I do a lot around the house. For one thing, I do all those traditional manly chores like shoveling the snow, mowing the lawn (at least until this year, when I found Michael, who does a much better job on my lawn than I do), most of the yard work (except for the flower beds, which are my wife's bailiwick), taking care of the cars, fixing any computer problems, and handling other electronic odds and ends. I try to put my dirty clothes where my wife wants them, and once in a while I actually run a load of laundry—although I have to admit a blue moon probably comes around more often. I *do* cook, and I make a regular effort to wash the dishes as well, or at least to pitch in and help my wife clean up in the kitchen.

Despite that impressive list, if I were in the confessional, I'd have to admit that I *don't* do fifty percent of the chores. Whatever I do get done, however, my wife appreciates. I, in turn, have the utmost appreciation for how much my wife does to keep our household functioning.

Appreciation can be shown by words. My wife might say, "Thanks so much for washing my car. It looks great!" Not only is she expressing appreciation for my effort but she's encouraging me to do it again.

Appreciation can also be shown by actions—not just the act of a thank-you kiss, although that's a good place to start. Pitch in and do the dishes after he's worked hard to make a great meal. Do as my wife did, even if it's only once, and hire a neighborhood kid to cut the lawn so your SO can nap in front of the game on Saturday afternoon or hang with his buddies.

Stop and Take a Moment to Appreciate What the Other Does

◆

I'm not the only one to suddenly realize my good fortune while pondering everything that my SO does to make my life and our relationship the best possible. One respondent said it very simply: "I just realized after filling this out that my husband takes care of me, the baby, and the household."

The Importance of Flexibility

My wife is also incredibly understanding about my work, which, since becoming a spokesperson for the Emily Post Institute, has consumed much more of my time—including time spent traveling for seminars, speeches, and appearances.

My own changed situation clearly illustrates one of the most important aspects to keep in mind when considering how to divide up the chores: Nothing in life is static. And when life situations change, you also need to review the dynamic of who's doing what. The hiring of Michael, who now mows our lawn, is a perfect example of this. In spite of the fact that mowing was "my" chore, my wife recognized that given the reduced amount of time I now had at home, the time I was spending mowing meant that I wasn't getting time to relax—by playing a round of golf, for instance. So she suggested Michael. "You can't do it all, and you need time for yourself," she said.

As I think about it, I realize what my wife did was etiquette personified: considerate, respectful, understanding, and acting with initiative to make both of our lives better.

ETIQUETTE IMPERATIVE

Make a List

───────◆───────

Instead of always thinking about all the things that *you* do around the house and yard, take a moment to catalog in your mind all the things that your SO does. Then take some time to reflect on and truly appreciate your SO's contributions.

Survey Says: It's All a Matter of Perception!

How do men's and women's perceptions match up regarding how much effort each partner puts into various chores? To answer this, the couples survey asked respondents to rate seventeen specific chores on a five-point scale, indicating whether the respondent typically did the chore, the couple shared the chore equally, or the person's SO usually did it.

When all the household chores are taken together as a whole, men and women agree: Women do more, by about a margin of fifty-five to forty-five. Interestingly, though, men's and women's opinions can sometimes differ when it comes to specific chores. Read on.

The Bathroom

This is my favorite room in the battle of the sexes. Consider for a moment all of the many items your bathroom contains, and where they're kept: If you're like most male–female households, she has laid claim to every surface, while every drawer (except perhaps half of one) is stuffed with her things; he gets that half-drawer. Clearly, this is *her* room. And she wants it kept a specific way. As a result, according to the couples survey, the woman cleans the bathroom most of the time. Sometimes the man helps out. But savvy male significant others will never risk moving any of her "things," for fear of putting them in the wrong place—and then not being able to tell her where they are, because he has no idea *what* they are.

In return, it is imperative that the man *always leave the bathroom the way he found it.* If she is caring for it, he has to respect that effort. So he always remembers to wipe down the counter after shaving or using the sink, he promptly removes any hairy mess from the shower drain, he hangs up his towels (and hers, if they are left out) immediately after showering, he cleans off the toilet rim or seat, if needed, and, yes, he puts down the toilet seat.

Meals

From the shopping to the cooking to the cleaning up, when it comes to the chores involved in meal preparation, men clearly give credit where credit is due. Shopping in particular falls primarily to women, although there's also a strong minority of males who take the lead in this task. I can only attest to this anecdotally: Whenever I'm at the grocery store, I see a number of men there doing the shopping. In our house, shopping tends to fall to whichever person has less to do at work that day, or whoever's heading home first.

Cooking is recognized by both genders as an area where men do step up to the plate. In fact, numerous female respondents wrote of how appreciative they are that the men in their homes do all of the cooking! Now, this arrangement may be a function of job demands; it may reflect the fact that the woman of the house hates to cook, or it may simply be that the man enjoys cooking and wants to prepare all the meals. Whatever their motivation, today's couples are steadily breaking down the stereotype of the woman as cook.

The dishes are an interesting case. You'd think that, in general, the deal would be that if she cooks, he washes the dishes—

but that doesn't happen. This uneven split is one of the reasons why women are perceived as doing more than fifty percent of the chores. Not only do women cook but they also clean up, or at least share that chore. Men who cook, on the other hand, are more likely to let their SO do the dishwashing.

Of course, the best approach to cooking is to clean up as you go. The other night, I started cleaning up since my wife had cooked a wonderful chicken curry. I placed the dishes in the dishwasher and reached for the pots and pans. There were just two—one for the curry, and one for the rice. "Did you clean up as you prepared the dinner?" I asked innocently.

"Look at all the clean stuff drying there," she said pointedly. My powers of observation are well known in our house: Not only had she cooked, but everything she'd used in prepping the meal was washed, and all the ingredients and spices were put away. I knew right away that it was time for some extra appreciation. "You really are wonderful, honey!" Kiss, kiss, kiss.

Non-meal Kitchen Clutter and Cleaning the Bedroom

Here are two areas where men's and women's perceptions vary sharply regarding who does what: Over fifty percent of men report that they share in the job of keeping the kitchen and bedroom areas neat, while just twenty-seven percent of women say the kitchen is a shared task, and only twenty-nine percent of women identify the bedroom as a chore done together as couple. Instead, the large majority of women tell us they clean the kitchen and bedroom entirely or mostly on their own. Clearly, the typical man believes he's really sharing in these tasks, while the typical woman thinks she's doing it all. Who knows who's really right? If what you're doing seems to work, just keep doing it.

Test Your Chore IQ

———————◆———————

How do you divide up the chores and work at home? Try completing the following questionnaire separately and then go over your answers with your spouse. Where you agree, celebrate. Where you disagree, it's time to talk.

	SELF DOES ALL	SELF DOES MOST, SO HELPS	SHARE EVENLY	SO DOES MOST, SELF HELPS	SO DOES ALL
Cleaning up bedroom	O	O	O	O	O
Picking up clothes and putting them in hamper	O	O	O	O	O
Putting clothes in washing machine	O	O	O	O	O
Hanging clothes to dry or putting them in dryer	O	O	O	O	O
Folding clothes	O	O	O	O	O
Putting clean, folded clothes away	O	O	O	O	O
Cleaning up bathroom	O	O	O	O	O
Cleaning out storage areas (garage/attic)	O	O	O	O	O
Shopping for groceries	O	O	O	O	O
Cooking the meal	O	O	O	O	O
Cleaning up after meal	O	O	O	O	O
Cleaning up non-meal kitchen clutter	O	O	O	O	O
Maintaining cars	O	O	O	O	O
Doing yard work	O	O	O	O	O
Fixing broken items (faucets, windows, etc.)	O	O	O	O	O
Paying monthly bills	O	O	O	O	O
Taking care of children	O	O	O	O	O

Clothes

Men are forever being vilified for not picking up their clothes. Yet the couples survey indicates that seventy-one percent of the men polled believe they share equally in this task, and that fifty percent of the women agree that it's a shared task. While perception may be part of the issue here, the reality is that the task is getting done much more efficiently than the pundits would have us believe.

The washing and drying of clothes, however, are clearly tasks done predominantly by women. Folding the clean clothes, on the other hand, is one of the few traditionally female chores that a significant number of men claim credit for. Over half (fifty-six percent) of the men surveyed say they either share in or do most of the folding. When it comes to putting the clothes away, however, both men and women agree this is either a shared task or a chore that the woman does.

Attics and Garages

Our survey found that in most households, the woman gives way here. Attics and garages are viewed more as the man's domain, and so women also see these as being his areas to take care of.

Cars, Yard Work, and Fixing Things

The survey results are clear: The overwhelming majority of couples consider these areas to be the man's responsibility.

Bill Paying

This category is an evenly mixed bag. Thirty percent of both men and women report that they are the only ones who pay the household bills. The other seventy percent of either sex say they share the chore with their partner.

The Bottom Line

When all is said and done, two facts emerge clearly: First, every couple has to come up with their own unique solution for who does what and when—ideally, one that works for both partners. Second, all successful couples seem to find a happy medium—a point where neither person feels he or she is being unfairly taken advantage of. If you're both comfortable with your divvying up of the chores, great! If not, then it's time to start talking.

11

LEISURE TIME

THE CHORES ARE ALL DONE (FOR THE MOMENT, AT LEAST), there are no kids to be tucked away, and no meal to prepare or clean up after. Now, you and your SO have some time to do what you really *like* doing. When it comes to how you and your SO spend your spare time, the choice rests with the two of you. That also means, however, that it's up to you to make good choices in your leisure activities—choices that will enhance your lives and your relationship.

The Importance of Doing Things Together

Your first decision, of course, is whether to recreate separately or as a couple. In any relationship, it is important for each partner to reserve time just for him- or herself (see "The Importance of

Beware Playing Coach

◆

You may be a wonderfully skillful skier or tennis player or golfer or rock climber. But no matter how well you know your chosen sport, be very, *very* careful about coaching your SO. For both coach and pupil, getting better at any sport requires time, patience, and a willingness to suffer through a lot of mistakes—demands that can tax even the most understanding of relationships. Unless you're able to summon up an enormous amount of tact, you run the risk that your advice may turn your SO off from the activity forever.

A much better solution is to find a nonrelated coach to teach your SO. Your role can then be to sit back and enjoy watching the practice session, offering comments only when you're specifically asked for suggestions.

Doing Things Alone," page 126). But it is also extremely important to do things together. Being a couple is about sharing experiences; it's about enjoying each other's company and doing things together that you both enjoy. After all, if you have no shared interests, nothing in common to talk about, then what basis is there for the relationship?

I'm not necessarily talking about splurging on dinner and a show. The activity you do together could be something as simple as going for a walk. The key is to find something that both of you can do *and* that you both enjoy. I know numerous couples who ski together, play golf together, garden together. These types of shared activities typically develop in one of two ways: Either the two people share a common interest from the start or one partner makes a concerted effort to become interested in the other's avocation.

Case in point: When my brother got married, his wife Maureen took up golf. She struggled at first, like every new golfer. But one of the wonders of golf is that people of vastly different abilities can go out and play together and have a good time. Even though my brother is an excellent golfer, he was able to enjoy playing with his wife right from the start. Today, Maureen has evolved into quite an avid and accomplished player herself. Had she not been willing to take the plunge and try golf in the first place, however, she might have missed out on an activity that is now a real source of fun for her—and she definitely wouldn't have spent nearly as much time with my brother.

Several years ago, my wife started taking yoga classes at a local fitness center. What impressed me most about her commitment to the classes was that two of them started at 6:00 A.M.—and one of those mornings was a Monday.

Now, to know my wife is to understand that morning is not her best time. Years ago, I started bringing her a cup of coffee in bed to help her get up in the morning. So the first time she set the alarm for 5:20, I just chuckled to myself. Well, I was wrong: She was up, out, and off to the class without missing a step. And she kept on getting up, right through the Vermont winter—a feat that's particularly impressive when it's ten degrees or colder outside and there's still an hour and a half to go before the sun rises.

The following summer, Dana, the yoga instructor, offered a "yoga for golfers" class that met on Thursday mornings, also at 6:00 A.M., at our golf club. My wife persuaded me to try the class for several weeks. We went together, and I got hooked. Now, at least once a week, and sometimes several times a week, my wife and I go to yoga class. It's become something we do together: We support each other, encourage each other, and even excuse each other on those days when it simply isn't going to happen.

For me to become a yoga student, two things had to happen. First, my wife had to have an interest that I wasn't a part of, and she had to be willing to share this interest with me. Second, I had to be willing to try something I never would have tried on my own. Her willingness to share and my willingness to accept her offer were what made the difference.

ETIQUETTE IMPERATIVE

Make an Effort to Do Things Together

◆

To share more time with your partner, work on developing common interests that you can get involved in together. This includes being willing to try new things as a couple.

The Importance of Doing Things Alone

As much as I believe that partners should do things together, I also firmly believe that they should do things on their own. The couple that spends every waking moment together will soon find it hard to bring anything new to the relationship. I know from personal experience that it adds spice to our relationship when my wife tells me about something she has done, or heard, or is thinking about.

The trick is to combine both solo and joint activities in a balanced way that enhances your relationship. For example, I know numerous golf widows (and some golf widowers) who simply don't care to play the game themselves. Still, they absolutely encourage their partners to play. In addition, they show up at social events associated with golf, attend the pre- and post-tournament dinners, and take an interest in how their partner's game was that weekend. They appreciate their partner's passion—but they never approach that first tee.

When my children were growing up, we often went skiing as a family. To my wife's credit, when the kids were young, she came along, too. Over time, though, she came to realize that downhill skiing really wasn't something she enjoyed doing. So when our daughters reached an age when they could fend for themselves on the slopes, she begged off one day, and then the next and the next. When we rejoined her at afternoon's end, she'd be very interested in how our day had gone and how the skiing was, but she far preferred hearing about it afterward to participating. Meanwhile, she got some well-deserved (and probably much-needed) private time to herself—which made her that much happier when we were all reunited.

The Importance of Doing Things Together Alone

While I'm a big proponent of family time and doing things with your children, I am also a strong believer that spending time alone with your SO—just the two of you—is vital to the success of any relationship, for two reasons:

1. Time alone with your SO helps you connect with each other and keep alive the spark that brought you together in the first place.

2. A little time spent *apart* from children, friends, and extended family will leave you looking forward that much more to spending time *with* them.

Your Special Place

It could be a particular bench in the park, a restaurant tucked away in a tiny alley, or a bed-and-breakfast in the country. It could be a roller coaster ride, or a pedestrian walkway on the Brooklyn Bridge, or a sandy beach in the Florida Keys. These are all settings that are simply another place for most people, but for you and your partner, they're imbued with meaning, either because something special happened there or because they are chockful of the very best of memories. Every couple has these spots, and every couple should make the effort to treasure them and to revisit them. They are your special places—and they help define who you are as a couple.

Who's Going to Watch the Kids?

◆

Babysitters are expensive, even when you're just going out for the evening. If you plan to go away for a long weekend or a vacation and leave the kids behind, then you're talking about serious money for child care. If the cost of hiring a babysitter is hurting your pocketbook, consider asking your relatives to watch your children from time to time. I know several families in which the grandparents relish the opportunity to take the kids for a couple of days. Another option is to arrange a kid swap with friends who have children around the same age—a sort of "you scratch my back and I'll scratch yours" approach. Do whatever it takes, but find *some* way to get off on your own. Your relationship will be that much the better for it.

My wife and I travel to Italy frequently—every year, if possible. We've done a lot of different things in the course of our visits there; we've seen many beautiful spots, met lots of interesting people, and found spectacular places to eat. As anyone who's visited Italy knows, while you may go there to see the sights, you quickly discover that the food is the real reason to keep returning to Italy again and again. We have our own favorite restaurant—well, actually two. Both are in Rome. One is a small local restaurant, Lagana, that we stumbled on the first evening of one of our vacations. Now, every time we visit Italy, we go back there religiously for our first meal of the trip. The food is wonderful, especially on that first evening in Rome following the long flight across the Atlantic. The antipasti are outrageously good—plates piled high with every kind of seafood and roasted vegetable, mozzarella, prosciutto and other meats, melon, steamed mussels, and sometimes even raw oysters. It's a true feast. The sliced steak, covered with olive oil on a bed of fresh arugula, is perfection.

For the final meal of each trip, we visit Ristorante San Eustachio, located near the Pantheon. Tables are set out on the sidewalk and the street, as well as inside. We try for an outside table whenever possible. With the warm night air brushing against us and the sights and sounds of people walking by providing a backdrop for our conversation, it's the perfect way to spend our last night on vacation.

These are our special places—places for which we have fond personal memories, places that make us feel good, places that allow us to feel particularly warm toward each other and to focus on each other. These places aren't associated with any holiday or time of the year; they're special to us simply because we discovered them together and they are the setting for some of our greatest memories.

Date Night, Part 1

For Jill and Steve, Thursday evening is sacrosanct. It's *their* night. Sometimes they go to the movies and get a quick bite to eat afterward. Other times, they might decide to paint the town—in which case they get dressed up and go to a great restaurant and then a nightclub. Or they may simply go parking on a back country road, with a few favorite CDs as background music.

Their Thursday-night rendezvous allows Jill and Steve to do all sorts of fun things, including some activities they never really thought they'd ever try—like the night they went bowling. Even more important, it also gives them a chance to concentrate just on each other for one night out of every week. Their date night is a chance to reconnect, reaffirm their relationship, and perhaps learn something new about each other.

What makes Jill and Steve's date night work is the fact that they talk about the evening in advance. They make a plan, and then they follow through. For them, Thursday night is a priority—something they anticipate and look forward to all week long. They know that their special time together is a key to helping them build a strong, lasting relationship.

Date night matters most when you're raising children. It's far too easy to let the children become an excuse for not spending any time alone with your SO. When you skimp on this alone time, not only does your marriage suffer but your relationship with your children suffers as well. As a couple, you need that time together to recharge your batteries so you can return to your family refreshed and renewed.

The most effective date nights . . .

+ Are sacrosanct, and are planned on a recurring schedule. They might occur on the third Saturday of each month, for instance, or every other Thursday night. Pick whatever works for you, then stick with it.

+ Have a little variety to them—dinner and a movie one night, followed by a ball game or a symphony concert on the green on a summer's evening.

+ Are occasions for you both to put on the dog just a bit, by showering, putting on a favorite perfume or cologne, dressing one notch up, and trying those fancy shoes you never wear.

+ Have a rule that says work—no matter how pressing— is not allowed to interfere with your evening's plans.

Date Night, Part 2

Planning your evening together as a couple is only half the fun. The other half is actually going out. And on this night, manners really *do* matter.

This is a time to focus on each other: Holding doors, holding chairs, holding her coat—these things underscore the special-ness of your partner and of the evening. So does saying yes—and adding *please*—when he asks if he can open your car door. So do saying "thank you," focusing on each other, linking your arm in his or hers, and not taking your meal for granted but rather enthusing about how good the pâté and the duck à l'orange are. These respectful touches all make a difference in elevating the night into a memorable bonding experience. (Of course, none of these things will matter if they're done only because you have to, and not because you want to. Remember—sincerity matters.)

Here are a few other things, large and small, that you can each do to make the night special for your SO:

+ Be on time to start the evening.

+ Have a smile on your face.

+ Leave your work and your troubles behind for the night.

+ Take a break from chewing over the same old same old (kids, work, etc.) by coming prepared with some interesting topics that you'd like to talk about with your SO. Look at the newspapers, read *People* magazine, catch up with ESPN, or make a point of finding out the latest news regarding your SO's favorite hobby or sport or activity.

+ Make a pact not to let the conversation turn into a grouse session about Aunt Tilda or about the Browns who live just down the street but have never invited you over.

+ Agree that if you bump into friends, you'll say hi but you won't join them or ask them to join you. (You can tell them why then and there, or call the next day and explain if you feel you have to.)

✦ At the end of the evening, take a moment before you go inside the house to thank each other for the great time you had together. Maybe even rekindle the magic of a good-night kiss on the front doorstep.

✦ If you really want to cap the evening off, send a thank-you note the next day to your SO. Keep it simple, short, and sweet. It's a great way to relive that warm feeling all over again.

The Power of Surprises

Etiquette is about enhancing relationships. The status quo of a good relationship is always comforting—but every now and then, a pleasant curve ball is in order.

Surprises come in all forms. They can be a gift, large or small. They can come on a special day: a birthday, an anniversary, or at holiday time. These types of gifts are somewhat expected, although the largesse of the gift can still make it a surprise:

"A new speaker system. This is awesome!"

"Diamond earrings! I really wasn't expecting them."

"A plasma television!" (Need I say more?)

Of course, a surprise gift doesn't have to coincide with any particular day, nor does it have to be pricey; the present could be something as simple as flowers or a pint of Ben & Jerry's ice cream. Doug told me how he stopped to get groceries one night on the way home from work and discovered that his grocery store discount card had accumulated enough points to earn a free greeting card. So he took a moment to read the cards—the ones with a general message, not the birthday cards—found one he liked, and took it home. Instead of simply handing it to his wife as they emptied the grocery bags, he signed it and placed it on her pillow. Good move.

Know When to Go Easy on Your SO

When you are planning your social calendar, the art of compromising and not complaining goes hand in hand with the art of knowing when your SO really doesn't feel like doing a particular activity. We each need to be sensitive to how our SO is feeling, and to support him or her when a hard choice has to be made.

More than once, I've come home to the sounds of my wife cooking up a storm in the kitchen. Candles are on the dining-room table, and it's been set with the nice plates and glasses instead of the everyday stuff. Soon, she calls me to dinner—and what a dinner! It might be veal scallopini with asparagus and noodles. Or maybe she's made a wonderful, warming butternut squash soup. It's such a simple surprise, really, but boy, does it make me feel wonderful.

Whether the surprise is big or small, the fun lies in planning it and carrying it out, and then seeing the reaction. There's nothing better than knowing you've made the day just a little brighter for the person who matters to you most.

Negotiating Your Social Schedule

Trying to do things together can get tricky when schedules and interests collide. Attending the University of Vermont men's ice hockey games is a perfect example. I've been going to their home games—and some of their away games—for over twenty years. Now, The Gut—UVM's Gutterson Fieldhouse—is not the most comfortable place to watch a hockey contest. We're talking hard wooden benches with no backs. My wife can handle two periods, but that third period is an absolute killer. And going with me to a hockey game and expecting to leave two-thirds of the way through just isn't going to work. So she doesn't go.

The problem is, the home games are played on Friday and Saturday nights.

Here's what we do: At the start of the season, we mark out all the UVM hockey games on the calendar that hangs in our kitchen. This calendar runs our lives. If an event is written down, it's going to happen—well, most of the time. I'm not foolish enough to think that for six or seven weekends each winter, we're simply not going to attend any social events. Instead, we both make some compromises. My wife tries hard not to schedule social activities on home-game weekends. Still, every now and then an invitation comes up that she really wants to accept. And that's when I have to compromise, by giving up a game and going with her.

Besides being willing to compromise, there's one other thing we both agree on—in fact, it is the thing that maintains our sanity: *We don't complain about the compromises we've made.* We each know we're going to have to give in on some things: There will be some invitations to events that won't get accepted and some events that my wife will go to on her own. And there are also clearly going to be some evenings when I won't attend a hockey game. But neither of us makes a stink about these compromises. Instead, we make the choice and then go with it. I'll attend the party and have a good time. Or if we skip a party, my wife will not say a word about what she's missing out on.

Funny, too, how when you don't hear any complaining from your SO about the last time, it becomes that much easier for you to be the one who compromises the next time.

A Night at the Opera

One winter afternoon a couple of years ago, my brother called us. He and Maureen drive up to Montreal several times each year to attend the opera, but this particular week the people they usually went with couldn't go. Did we want the tickets?

My wife had answered the phone. She was definitely interested, she replied, but she'd have to check with me. Later, when she did talk to me about it, she did it in such a way that I could tell this was something she really wanted to do.

Now, I had never been to an opera before—and to be honest, the few times I'd caught a performance on NPR, I had not been overwhelmed. I'd been told that opera was one of those things you either liked or you didn't, and I'd always assumed that I fell in the latter category. In my personal universe, going to the opera was something of a low priority, to say the least.

But it was winter in Vermont, when cabin fever runs rampant. And there was no hockey game that weekend. And she had that look on her face. "Okay, let's do it," I said. After all, how bad could a weekend in Montreal be? A night away from home, a good dinner at a nice restaurant . . . and the opera.

One of the most important things I teach people in my etiquette seminars is how to recover from mistakes. When you've made a mistake or have clearly been wrong about something, the first thing to do is to own up to it. When I walked out of the opera that night, I had to admit that I truly had enjoyed the performance. In fact, it was so pleasant that my daughter gave us opera tickets for Christmas the following year, and we went back a second time.

Leisure Time at Home

Of course, you don't necessarily have to go out every time you have a spare moment. You can also enjoy each other's company at home.

One of the simplest things we do is watch television. Or, rather, I watch television. My wife, on the other hand, is an avid reader. Even though I know she'd prefer to read in a quieter setting, she will come into the room where I'm sitting, settle down next to me with her book in hand, and start reading. She could easily read in another room or snuggle in bed, but instead she chooses to be with me. When I think of gestures that help couples to bond (see Chapter 5, "Thoughtful Gestures Make All the Difference," page 55), I think of things like her simple action of sitting with me.

Your leisure time together at home is also an opportunity for surprises and unique experiences that can add spice to your everyday lives. You never know when these opportunities are going to present themselves, so be ready to jump when one appears. Case in point: One evening, my wife and I went to a charity auction for a local theater. One of the items on the auction block was an evening with a local chef who would come to your home, prepare a meal for you, and clean up afterward. What a great way to spend an evening with my wife, I thought. The bidding started and, after raising my hand several times, I suddenly found that everyone else had dropped out and we had won ourselves a chef for the night.

Checking in With Your SO

◆

When my brother called to invite us to the opera, my wife told him she had to check with me before accepting—even though she was absolutely dying to go. No matter how tempting an invitation sounds, never commit to any social event without first checking with your SO. You should do this out of respect for your partner, and also for the practical reason that there may be a conflicting event on that date that you don't know about.

The dinner that evening was a lot of fun, and quite different from anything we would have cooked for ourselves. And for one autumn evening, we didn't have to do a thing except enjoy ourselves and each other.

The Art of Compromise: It Won't Always Be Fifty-Fifty

"Tonight we did it your way, so tomorrow night, we watch what I want to watch." This even-steven approach might work with television viewing, but in general, it's hard to ensure that all of your leisure time compromises will come out even.

Over the course of the hockey season, there will invariably be more games that we compromise on and I go to than there will be games I miss. How do you equate a hockey game that results in a missed party with a decision to skip a game and go to the opera in Montreal?

Answer: You don't.

The minute you start counting and comparing how many times *she's* compromised to the number of times *he* has, you're heading down a slippery slope. Similarly, if you start rating the quality of the compromises, you're also heading for big trouble. What counts is that over time, both parties feel that they are enjoying their lives. They have to feel that they are each doing things on their own and aren't being unfairly prevented from doing those things; that they are also sharing things together that they both enjoy; and that occasionally, each partner is stepping outside his or her own comfort zone and trying something new—both for the sake of trying it and to share the experience with the other person.

12

WHEN REASONABLE
PEOPLE DISAGREE

IT'S NOT A MATTER OF IF, BUT RATHER A MATTER OF WHEN. Even the most compatible couples have disagreements, and even the most reasonable of partners will find themselves at logger-heads from time to time.

My experience is that in personal situations, most disputes arise when one partner makes a misstep and the other partner reacts badly:

- ✦ He snaps at her for leaving the car empty of gas.

- ✦ She fumes at him because she tried to use their joint credit card and it was declined.

- He snipes at her because she asked her mother to baby-sit the kids yet again, instead of asking *his* mother this time.

- She lashes back at him because he never arranges the babysitting—and maybe if he did once in a while, his mother would get asked more often.

- He growls at her because she isn't ready to leave at the time they'd agreed on.

- She points at him because he didn't call to say he was going to be late getting home from work.

The list goes on and on. One person has had a tough day for whatever reason, the other does something that wasn't quite as thoughtful as it ought to have been, and suddenly cross words are spoken, feelings are hurt, and everyone ends up feeling miserable.

Living peaceably with another person really comes down to two things: *compromising* and *accepting responsibility* when things go wrong.

Pick Your Battles Carefully

Thirty-two years of marriage, and my wife and I are still two people living together. Of course we've had our disagreements. We each do what you might call foolish things from time to time, which naturally elicits a less-than-pleased reaction. We've both also learned over the years that some behaviors simply aren't going to change, no matter how often one of us brings them up to the other.

Consider Roz and Sam. At night, Roz is routinely the last one to climb the stairs, and when she does, she invariably leaves a couple of lights on downstairs. Sam is reminded of this oversight every month when he pays the electric bill. Each time Sam mentions the issue, Roz makes an effort to remember to turn out all the lights, and for a while everything is okay. But eventually, she relapses and the problem raises its ugly head yet again. Now Roz's blind spot could easily be fuel for heated words between them. In fact, the last time Sam talked to Roz about it, the conversation had turned unpleasant. Sam, to his credit, now realizes that reminding Roz over and over again isn't getting them anywhere as far as permanently resolving the problem, and that saving a few dollars a month isn't worth the hurt to their relationship. And so he's decided to stop making a big deal out of the problem.

Good choice.

Bridges You Don't Want to Cross

In any relationship, certain topics are better left alone. For instance, you and your SO may have different views on politics. If this is the case, during election season you may well find that the best course of action is to avoid discussing the pros and cons of each person's favorite candidate. It's difficult enough to carry on a conversation with a friend who backs the other side without letting the banter turn confrontational or personal. If a similar conversation occurs with your SO, things could get said that would be very hard to take back.

One danger sign of any discussion is when the talk turns from comments about the issue to unfair characterizations of the person you are talking with. You've crossed this bridge when the talk gets personal: "I can't believe you're so far to the right of Louis the Fourteenth that you actually agree with that idiot" versus "Joe Candidate's stands on the issues all seem to be very conservative. Are you comfortable with that?"

If you do find yourself in a spirited debate with your SO—whatever the subject might be—always focus on the issue at hand, and never on each other's personality.

When Mount Anger Erupts

It happens to everyone. Pressures can build up, for all kinds of reasons. And then all it takes is a spark to ignite an eruption. Two things to note about blowups:

1. Sometimes they can be a good thing. If a blowup is over a legitimate issue, clearing the air can be a step toward resolving the problem. At the same time, for the angry person's partner, a blowup is a moment that tests his or her commitment and love, since the partner often is the object of the blowup.

2. The best response to a blowup is to disengage and wait until the other person has vented and calmed down. It is very difficult to reason or compromise with anyone, much less your SO, when that person is agitated. In the heat of the moment, your partner's responses are likely to be either attacks on you or outright denials of the situation. Return to the discussion after both of you have had time to detach from your emotions and are able to focus just on the issue at hand. This will give you a much better chance of resolving the conflict.

Not Always Fifty-Fifty, Revisited

Jill and Barry are redecorating their living room. Jill favors a soft blue color for the walls, while Barry likes yellow. Clearly, they're not going to be able to halve the difference on this one. A fifty-

Anger Is Better Than Apathy

◆

While no one enjoys a conflict, it's actually a sign of an engaged relationship when a couple disagrees about something and expresses this disagreement to each other—even if one or both partners are angry. As long as the two of you are communicating, you have the ability to deal constructively with any emotions you're both feeling. A far worse situation is when one or both partners shut down emotionally and can't be bothered to express their feelings or opinions. Apathy is a danger sign, indicating that a person is disengaging from a relationship. It signals that he or she has crossed a critical point and simply doesn't care anymore.

Give me anger over apathy anytime. When one partner is apathetic, resolving any conflict becomes much more difficult, because it depends first on that person somehow becoming engaged again. But whenever there are real feelings being communicated, including anger, there is always an opportunity to resolve the issue.

fifty split would have them painting two walls yellow and two walls blue—which isn't going to happen. Therefore, either Jill or Barry is going to have to compromise and accept the other person's choice, or they're going to have to find a third color they both agree on.

Typically, successful couples end up resolving this type of issue by deferring to the partner to whom the issue matters more. Men I've talked to say they tend to leave decisions related to decorating in their SO's hands. So in this case, Barry defers and says, "Blue is fine with me." Most couples survey respondents recognized that compromise is not about splitting every choice down the middle, but rather about having each person's choices honored and respected—even if that choice isn't the one the couple ultimately settles on—so that each person feels his or her desires are being taken into account and fairly acted on:

"My wife and I respect and trust each other's decisions, because we know which one of us is the strongest in the area of the particular issue being decided."

"We realized early on that compromise does not necessarily mean meeting in the middle. We used to 'settle' on movies that we both sort of wanted to see. Now, we switch off so that each one of us gets a chance to see our first choice. We also take 'split vacations.' For example, this year we're going to New Orleans for me, and then to Orlando for her. And for each other's sake (as well as our own), we both put in a sincere effort to enjoy the part of the trip that is not ours."

Don't Let Peas Turn Into Boulders

Have you ever become obsessed about some small issue in your relationship, until it starts to loom larger and larger? It's like the story of the princess and the pea. To the princess, the irritation caused by the tiny pea under her mattress was magnified until it felt like a boulder. In your relationship, the source of irritation could be a difference of opinion, or a habit of your SO that bothers you, or a thoughtless comment you can't seem to get out of your mind. Whatever the issue might be, however, the real problem is that it never gets addressed. Instead, it remains stuck in the fabric of your relationship, like the pea under the princess's mattress, and the irritation festers and grows until it is way out of proportion to the original issue.

The bottom line: If something in your relationship bothers you, don't just let it slide. Remember another one of Murphy's Laws: Left to themselves, things tend to go from bad to worse. If you think that your SO's troublesome habit will simply vanish on its own, you are mistaken. The earlier any problem is addressed, the easier it is to correct—and the later you wait, the harder it will be to fix whatever is wrong.

Someone's Looking for a Fight

———————◆———————

You can see it in the other person's eyes and body language long before the words start coming out: Your SO is looking for a fight, and you're the target. So what do you do?

Fortunately, there are two keys to defusing the situation: (1) *engage* the other person *very gently*, and (2) *don't belittle* the other person for his or her mood.

If you know what he's angry about, head him off at the pass by apologizing and offering a solution. "Tom, I can see you're upset, and I know it's about my not calling you when I knew I was going to be late. I'm sorry. I'm going to . . ."

If you don't know what the anger is about, you can still head her off at the pass. "Cindy, I don't know if it's something I've done or something else that happened, but you seem very upset. Can we talk about it?"

How to End It and Move On?

Making up is the last critical step in resolving a disagreement, especially for a couple in a relationship. Here's why: When you have a disagreement with a friend or a coworker, you go your separate ways afterward. That separation helps you both to let go of your emotions and get over the problem.

At home with your SO, though, there *is* no separation. Sure, you can sit in separate rooms for a time, but eventually—unless you are one of the few couples with separate bedrooms—you're going to have to climb into bed together.

One of the most common themes mentioned by successful couples is that they have developed some way of breaking through the temporary emotional barrier that goes up after a heated argument—some kind of private signal that says, "I'm sorry we argued, let's make up."

"We stick our tongues out at each other," one focus-group participant offered. For her and her SO, the silliness of a tongue sticking out breaks the tension and creates an opportunity for laughter to come in—and laughter is one of the most powerful antidotes to emotional tension.

Another participant pointed out the powerful force that simply saying "I love you" can have in quickly dissipating the tension and bringing a swift, positive resolution to the disagreement.

Similarly, one of the most effective things one partner can do is own up to the problem directly: "Tom, I'm sorry. You're right—I should have called. And I'll try to do better the next time."

It is tough to have an argument with a person when there's nothing left to argue about. Remember, it takes two people to have a dispute. By accepting the responsibility when it's legitimately yours, instead of rationalizing or explaining away your mistake, you move beyond the argument. Almost invariably, what the aggrieved party really wants is for you to own up to what you did, show that you understand it caused the other person difficulty, and offer some form of assurance that you'll try not to repeat the action.

Four Steps to Resolving Disagreements

Here are four checkpoints that will help bring any dispute to resolution:

> **Step 1: Disengage and get calm.** When tempers are flaring and emotions are running hot, the participants are much more likely to get their backs up against the wall and refuse to budge from their positions. Result: stalemate.

> **Step 2: Stay clear of personal attacks.** Remain focused at all times on the topic of dispute. Veer into a personal attack, and you'll find yourself back at step 1.

Step 3: Articulate each person's position. It's easy to say what *you* believe, but a lot harder to articulate the *other* person's position. When you do so, however, you'll have more insight into your partner's position, rationale, and concerns. This insight will help you develop a solution that works for both of you.

Step 4: Seek closure with a solution that's amenable to both partners. A solution that only one person agrees to isn't really a solution, and if you attempt to implement it you'll quickly find yourselves in another argument. Each person has to understand and agree to the solution—whether it involves finding common ground, giving and accepting an apology . . . or settling on blue paint instead of yellow.

ETIQUETTE IMPERATIVE

There's No Substitute for "I'm Sorry"

Apologies are important. They need to be explicit—"I'm sorry"—as opposed to implicit—"Yeah, well, I guess you're right."
And no *but*s!

PART TWO

The Two of You in the World

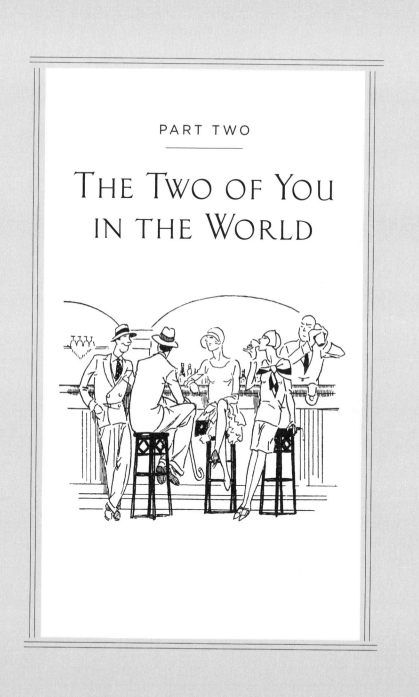

13

THE PUBLIC COUPLE

THE REALITY IS THAT WHEN YOU'RE PART OF A COUPLE, your actions, appearance, and words don't simply affect people's view of *you alone*. They reflect on your partner and on your relationship as well. When that reflection is positive, it bathes the two of you—as individuals and as a couple—in a wonderful light. If the reflection is negative, however, it casts a wide shadow. When you are thoughtless in your actions, appearance, or words, it reflects poorly on you as an individual, of course, but it also reflects badly on your partner and on the two of you as a couple.

The 24/7 Partner

Before Jeff heads out for his usual round of Saturday-morning errands, he does the right thing by asking Sarah if she needs anything. "I'm all set—but don't forget Jane and Dick Henderson have invited us to a dinner party tonight," she reminds him. "I know you haven't met him yet, but she's really nice, and I've been looking forward to getting together with them."

Jeff heads out with his list. First stop is the home center for some hooks and wire to hang the new pictures Sarah just bought. Purchase made, Jeff starts pulling out of the home center parking lot as the light turns green. Suddenly, a car cuts through the intersection, nearly hitting Jeff. His blood pressure skyrockets. He leans on his horn, hurls a few choice epithets directly at the other driver, and then caps off his frustration by flashing a well-known hand signal.

You guessed it: The other driver is Dick Henderson.

When Jeff and Sarah arrive for dinner at the Hendersons' that evening, Jeff and Dick recognize each other. Let's hope that apologies all around along with a good laugh can sweep this incident under the rug. At the very least, though, this thoughtless behavior—on both Dick's *and* Jeff's parts, to be fair—ended up causing embarrassment for everyone.

Let's take a closer look at where both men went wrong:

✦ **Dick:** Rushing, taking chances, being aggressive—these types of behavior all lead to trouble. Other drivers can easily become frustrated when confronted with drivers like Dick, and many will react instinctively with angry gestures, cross words, or (in this day and age) far worse. The newspapers are filled with too many stories of road rage run amok. Dick's choice to push the limit in going through a yellow-turned-red light not only endangered himself and others but it also caused Jane a sudden moment of real difficulty when Jeff and Sarah knocked on their door. Now, a potential friendship needs salvaging before it has even started. Not a good way to begin the evening!

✦ **Jeff:** His reaction escalated from a cautionary horn blow to a personal, vitriolic attack on Dick. Car horns are meant to be used, and Jeff's blowing of his was appropriate. When he started mouthing off and gesturing, however, he crossed the line from acting in a rational way to reacting thoughtlessly. On a practical level, when you confront a stranger, you have no way of predicting his reaction. Again, read the papers. There are stories of people chasing other people down after a confrontation and attacking them with fists, tire irons, even guns. You simply don't know how a stranger will respond when you turn an issue into a personal attack. Meanwhile, on the level of social friendship, Jeff has clearly placed Sarah in the same awkward situation that Dick placed Jane in.

Don't think for a minute, by the way, that it's only men who lose their temper while driving. At a recent seminar, I laid out this same scenario for the participants. All at once, a table of five women started laughing. They then told me that on the way to the seminar that morning, one of the women had found herself in a situation where *she* was the one to raise her finger in a frustrated salute to another driver. My guess is that almost every driver has been there and done that at least once in their driving career.

The point is, how you conduct yourself matters to your coupledom. This is true even when you're doing something as individual as running errands by yourself. You, your SO, and your relationship are all hurt when . . .

✦ Going down the aisle of the grocery store, you bump into another person's cart and don't apologize.

✦ You're talking to a clerk when your cell phone suddenly rings. You answer and begin talking, tying up not only the clerk but also the person waiting to talk to the clerk, who in turn is starting to fume. (I found myself in this position just the other day; I ended up leaving the store without buying anything.)

✦ You pull into the parking lot, see a space, and gun your car's engine to get there just in front of someone who was coming from the other direction with their blinker on. (This situation was actually the subject of a recent television advertisement extolling the capabilities of an SUV. I find it incredible that a car manufacturer would approvingly count rude behavior as a quality trait of the people who buy its cars.)

✦ You arrive at the checkout station, then make everyone wait while you run back for that one little item you forgot.

✦ You jaywalk coming out of the store, almost causing an accident as a driver swerves to miss you.

The list is endless. You can easily imagine how each of these situations reflects on the perpetrator. Unfortunately, the action may also cause difficulty for the perpetrator's SO, too. You simply never know when you and your partner are going to meet someone. If that someone was on the receiving end of poor behavior by your SO, the actions will affect you and your coupledom as well.

ETIQUETTE IMPERATIVE

Alone or Together,
Your Actions Reflect on Each Other

◆

Etiquette and manners matter to your relationship. This is true not just when you're with your SO but also whenever either of you is in contact with any other person.

Why Manners Matter, Revisited

Yvonne is walking down the street when she sees Maggie coming the other way. Yvonne's got a cold, so when Maggie extends her hand to greet Yvonne, Yvonne doesn't reciprocate. Maggie wonders: "What's going on here? Is she mad at me?"

At this moment, Yvonne has failed to follow the expected social custom, which is to offer her hand. As a result, instead of a typical greeting and a short, pleasant conversation, the start of their interaction has been marred by Yvonne's failure to do what's expected. To her credit, Yvonne suddenly realizes her faux pas and says, "I'm sorry I didn't shake hands. I'm coming down with a cold, and I didn't want to give it to you."

All is forgiven.

This story is a perfect example of manners in action.

+ **Manners tell us what to do.** Maggie approached Yvonne and extended her hand in greeting, a very usual and expected gesture.

+ **Manners tell us what to expect others to do.** When Maggie put her hand out, she fully expected Yvonne to do the same. When Yvonne didn't, the focus turned from the greeting to: What's the matter?

Anytime we don't do what's expected of us, it's important to acknowledge that variation in our behavior. Yvonne recovered by explaining that she had a cold. If, instead, she hadn't said anything, her faux pas could have reflected not only on her but on her husband as well. Here's how: Without Yvonne's explanation, Maggie might well have gone home and said to her husband, Craig, "That was really weird . . . I saw Yvonne at the store, but when I went to shake her hand, she didn't shake mine. I wonder what's up? I was going to ask them over for pizza and a movie this weekend, but now I'm wondering if maybe we should ask the Rollinses instead."

Dissing Doesn't Work

While basic manners matter, Post surveys repeatedly point out the frustration people feel when they aren't respected. Disrespect takes numerous forms, but in the end, it denigrates the victim. The following actions are the most commonly cited ways partners show disrespect—and unfortunately, they do it in public at least as often as they do it in the privacy of their homes.

Avoiding Putdowns

"Everyone knows Tom is the laziest guy in town!"

"Dinner at our place? Have you ever tried Jane's cooking?"

We can all wield a sharp-witted tongue from time to time, but when teasing banter crosses over the line into hurtful comments, it can wound a relationship. In general, think twice before ever throwing a clever barb that has your SO as the target. For one thing, you know his or her sore points all too well. And by zinging your partner—whether in public or in private—you are also implicitly saying, "Don't count on me to support you."

One indication that you've stepped over the line: When other people in the conversation hear an over-the-top comment, they'll usually say something. My daughter's favorite response to a particularly acidic comment is, "Ouch—that was harsh!" If you hear other people reacting negatively to the way you speak to or about your SO, don't brush them off. Instead, take their comments at face value and tone down your own.

Not Interrupting Each Other

When Jane's parents finally got to meet her new boyfriend, they couldn't help spotting a disconcerting trait. "Did you notice how every time Jane started to make a point in the conversation, Tom interrupted her?" said Jane's mother afterward. This habit is apparent to everyone (including the SO being cut off) except the person doing the interrupting. This disrespectful behavior can even persist when a couple is alone together. While we often don't even realize that we're doing it, interrupting demeans the person we are with. By interrupting his girlfriend, Tom is saying, "Jane's comment isn't worth hearing" – and, by inference, "Jane really has nothing of importance to say."

Ignoring Each Other

While interrupting is a sin of commission, ignoring your SO is a sin of omission. Tom and Jane are having dinner with four of Jane's oldest friends, and Jane devotes her entire attention to her pals. Throughout the evening, she barely talks with Tom and does absolutely nothing to draw him into the conversation. Result: Tom feels like the proverbial wallflower, and Jane's friends leave the restaurant without ever really having gotten to know him. There's no excuse for Jane's behavior. There's also little doubt that her actions will lead to hurt feelings and sharp words later.

Failing to Introduce Your SO

Introductions matter – and not just at parties, either. They matter when Jane and Tom bump into a business colleague of Jane's in the grocery store. Instead of introducing Tom, Jane acts as if he isn't there while she and her coworker start rehashing the details of the big deal they're working on. Next thing you know, Tom is memorizing the ingredients of the items on the shelf next to him.

An On-the-Spot Solution

Instead of letting these rough patches pass by without comment, Tom or Jane can take the bull by the horns and deal with the situation on the spot. The key to dealing with rude, disrespectful, or hurtful behavior is to *change the behavior* so you can end up having a good time – *not* to try and embarrass your partner. For example, when interrupted, Jane can stand her ground and say, "Tom, hang on a minute. Please let me finish my thought first. Thanks." If Tom is being ignored, he can change the dynamic by listening to the conversation for a bit, and then assertively making a comment: "You know, Sally makes a really interesting point. But I think she can take it even further by . . ." And whenever Tom is not introduced, he can introduce himself: "Hi, I'm Tom, Jane's boyfriend. You must be a friend of hers from Uptown Advertising."

Your Friends, My Friends, Our Friends

Being a couple, of course, encompasses much more than just your relationship with your SO. When the two of you become a couple, your coupledom affects both your friendships and your partner's. In addition, as a couple you begin to develop new friends—"our" friends, rather than "your" or "my" friends. How you interact with these various friends, both individually and as a couple, plays an important part in defining your relationship.

Friends from Before

Frankly, when we asked our couples survey respondents how well each partner got along with the other's friends, their answers surprised me. Anecdotally, we'd heard that precouple friends often had a hard time getting along with a new partner. But in our survey results . . .

+ Eighty percent of the respondents said that they get along very well or extremely well with their partner's friends.

+ Seventy-five percent said their partner got along very well or extremely well with their friends.

Compatible Personalities, Compatible Friends

One respondent in the couples survey on sharing friends offered this explanation: "We tend to share preferences in people, as in other things." Others pointed out that they'd moved in the same circle as singles, so naturally those same people continued to be their friends when they became a couple.

When both of you genuinely like each other's friends, it's a win-win situation for each of you and for your friendships. The following comments speak volumes about how relationships are enhanced when the two partners genuinely like each other's friends. Note the positive attitude shown toward friends and toward each other:

> *"I am more of a social butterfly than my husband, so I have more friends than he does, or shall I say I had more friends than he did—because, you see, 'my' friends have quickly become 'our' friends. My husband gets along so well with my friends, and vice versa, that if he answers the phone when a friend of mine calls, he/she will talk to my husband for twenty minutes before they ask to talk to me. I also really respect his friends, and enjoy spending time with them. It makes things really easy, because we never feel like we have to choose between our spouse or our friends. There is no tension when we have guests over for dinner because the group dynamic or chemistry is right."*

> *"I am very outgoing; my husband is very shy. But we like each other's friends just fine. We both must have good taste!"*

> *"The more time we spend with each other's friends, the more we like them!"*

ETIQUETTE IMPERATIVE

Don't Cut Those Old Friends Off

If your partner has friends you don't really cotton to, don't automatically turn the situation into one of "my way or the highway." Instead, try limiting your encounters to a level you're comfortable with, while also making an effort to see in the person what your SO sees. Remember, tolerance is the key to harmony in your relationship.

Dealing With Your Partner's "Difficult" Friends

Of course, since there's no accounting for taste in this world, the chances are reasonably good—no matter how compatible you and your SO's friends may be in general—that you will be less than enamored with at least one of his or her friends. Shawna and her husband Bill have developed a number of good friends as a couple, but Bill also has his friends from before they ever started dating. "A little rough around the edges" is how Shawna describes them. She knows that simply cutting Bill off from his old friends completely would not be good for their relationship, so she's made an effort to continue to see them as a couple—but she also makes sure that they do so only "in small doses." They see these friends occasionally rather than every weekend, for example, and always arrange to meet them at the movies rather than have dinner and an entire evening together.

Here's what some of our survey respondents had to say about getting along with their partner's difficult friends:

"It may take practice with some friends, but if they are important to your SO, then you should make every effort to become friends with them as well."

"Some of his lifelong friends are really wild and a little too much to handle—often to the point of being rude and

inappropriate. If I am in a situation that's uncomfortable, it's okay with my SO if I politely leave (I still want him to be able to hang out with his friends), or we will both check out early."

"On the rare occasion that one of my friends is not a favorite of my SO's, he still makes the best of it; he is friendly and thoughtful and engages them in conversation to make them feel welcome."

"We have very different friends, but we accept them for who they are, and we also accept them because of their friendship with us."

But what should you do when you truly don't like spending time with one or more of your partner's friends? "I dislike his friends immensely," wrote one respondent. "It's definitely an issue we have to work out." In this situation, compromise and commitment have to trump personal feelings, and each partner has to be able to communicate honestly about what he or she is feeling. Failure to work through this kind of conflict is sure to lead to major trouble.

Here are three possible solutions to the dilemma:

1.   Give your SO space and time to continue to see his friends by himself.

2.   Agree to see these friends at larger group activities, but limit their individual contact with the two of you as a couple.

3.   When spending time with a problematic friend, agree that when you've reached the saturation point, a signal from you will indicate that it's time for you and your SO to call it an evening.

Regardless of which approach you choose to take, the most important thing is that you communicate with your SO about the issue. Our survey respondents indicated repeatedly that when dealing with situations where one partner didn't like a friend of the other's, they were able to resolve the issue successfully by talking it over and establishing ground rules that were acceptable to both partners.

If the two of you don't communicate about problem friends,

The Immunity Rule

◆

When you connect with another person, you get the whole person—baggage and all. To warrant taking a stand over one of your partner's friends, that friend would have to say or do some incredibly egregious thing. One survey respondent put it best: There will always be at least one of those friends you just don't like. In dealing with this kind of friend, you have to weigh the pros and cons together very carefully. My friends and I have our own little rule: If your SO's friend was there before you, he or she has what almost amounts to immunity. Unless such friends are actually causing damage, either as a friend or to the relationship, we will go out of our way to compromise with them.

then one partner is left with the options of either suffering in silence or simply refusing to have anything to do with the other's friends—which is never a good thing for a relationship. "He only has casual friends. He is intolerable to my friends and is rude from time to time," wrote one survey respondent. "He doesn't get up from his chair to greet people or see them to the door when they leave. He yells good-bye from his chair." Besides being awkward for everyone involved, this attitude automatically creates an ongoing rift between you and your SO, which is a recipe for disaster.

Staying in Touch When You're in a New Relationship

You've fallen in love with a terrific person. That's wonderful—but be aware that some of your single friends may resent your new partner. Any new relationship inevitably alters the nature of the friendships you had previously. Suddenly, your evenings and weekends are taken up with your new partner. No longer are you available at the drop of a hat to grab a drink or go on a shopping spree. And with close friends of the opposite sex, you'll probably have to curtail your interactions even more.

As you work to successfully integrate both your old friends and your new partner into your current life, keep the following in mind:

✦ Don't lose touch with the people who were in your life before you found your SO. Make the effort to pick up the phone and arrange to meet them for lunch, or go to the movies one night, or get together for drinks after work—whatever works for your particular schedules.

✦ Make a point of inviting your old friends to the parties and dinners you host as a couple. This gives your friends a chance to discover why you think your SO is a genuinely great person and lets them see that just because you are in a serious relationship, you won't cut them off.

✦ See your friends from time to time without your partner in tow. It's very important for both of you to have friendships outside your relationship. The alternative—spending every waking moment with your SO—is neither emotionally healthy nor socially practical.

✦ Finally, realize that if an effort to integrate a certain friendship into your relationship should fail, you may have to make a choice. As one survey respondent wrote, "I put my friends on notice that this was the love of my life, and I hoped that they liked him — but if they didn't, then it was too damn bad."

Making New Friends Together

It was our first birthing class, and we were nervous. When we walked into the room, however, we were immediately surprised to see that, while we were just shy of thirty, most of the couples were much younger than that.

Class started. We watched a movie, and then came the question-and-answer session. A hand went up, and a male voice asked, "Tell me more about this water breaking. If it happens in my BMW, do I have to worry about the leather seat?"

I never did hear the answer, because I was too busy laughing at the question. I looked at my wife, and we both knew we had found another thirty-something kindred spirit. After the class was over, we introduced ourselves to Peter and Leigh and made plans to get together.

Our daughters were born ten days apart, and we have been the best of friends ever since. We take at least one trip together each year, and we socialize often, both as couples and individually. If Leigh is away, we'll invite Peter to dinner, and vice versa. We play golf together, and sometimes get in trouble together. (Once, when I was hospitalized briefly, Peter and a couple of other friends came to visit me at the hospital. He started recounting a particularly amusing adventure, and soon we were laughing out loud — the best medicine of all. That's when the nurse appeared at the door and threatened to throw us all out if we didn't quiet down.)

We also egg each other on to do the unexpected, as good friends will do. I have a pilot's license. Any pilot will tell you that any excuse to get in the plane and go somewhere is a good excuse. One day Peter got to talking about how nothing tasted better than Maine lobsters. The next thing I knew, while Leigh and my wife stayed home to prepare the rest of a summer dinner, Peter and I were flying from Burlington to Bar Harbor, Maine. We purchased fresh Maine lobsters from a roadside stand right outside the airport, then turned around and flew home in time for supper. What a great meal that was!

We also introduce each other to new people and new experiences. One night we met Leigh's brother and his new wife at dinner. That friendship has grown as well, both as part of our friendship with Leigh and Peter and in its own right. Recently, in fact, we spent ten days traveling in Italy with Mac and Virginia.

Friends enrich your lives and your relationship as a couple. It takes work and time to cultivate your friendships as a couple — but the result is a boon for everyone. One thing to keep in mind: Such friendships work best when both partners are comfortable with the other couple. "We have a deal that if we're going to be friends with another couple, we each have to like both members of the couple," notes one couples survey respondent. "If not, time spent with this couple will be limited."

ETIQUETTE IMPERATIVE

Friends Smooth the Edges

◆

Friends matter to a relationship, not only because of the enjoyment they bring, but also because spending time with friends takes pressure off you as a couple and helps put your relationship into perspective. A couple that doesn't bring friends into their daily life is left to focus only on each other—with the result that little issues can become magnified into big ones.

Boys'/Girls' Night Out

One thoughtful gesture you can make is to let your SO go off on his or her own from time to time, without feeling any guilt at leaving you alone. The classic version of this is the "boys' (or girls') night out." I've gone off with the guys on golf weekends or to a hockey game, and I have a number of male friends who enjoy a regular poker game. The night out with the guys happens occasionally, and when it happens, it's a lot of fun. My wife doesn't begrudge me these opportunities. I think we both believe that while we are a couple and we do a lot of things together, we're also two individuals who enjoy times with friends on our own.

That's why when my wife recently started having an occasional girls' night out with her own pals, I was all for it. After spending night after night in the constant company of yours truly, who could blame her for wanting to enjoy an evening out at a restaurant with her friends? Lately, they've started holding this "night out" at each other's homes. When it was my wife's turn to host the group, I entered the living room and pleasantly greeted everybody. Following that, I clearly recall hearing one of the friends asking if I wanted to join them. For about three seconds, I considered the offer. Then I wished them all a fun evening, skedaddled upstairs, closed the bedroom door, turned on the television for some background noise, and did several extremely productive hours of work.

14

OUT ON THE TOWN

WHILE THEY JOKE ABOUT HOW THEY SOMETIMES FEEL LIKE
hermits, Dawn and Tony actually do get out fairly often to enjoy
dinner with friends at a restaurant, catch a performance at the
local theater with another couple, or make up part of a table at a
gala charity event. And whether the activity is low-key or black-
tie, whenever they go out, Dawn and Tony aim to have a good
time. More than that, they want the people with them to have a
good time as well. After all, why else are they going out on the
town but to enjoy themselves?

As a couple, your best guideline to etiquette when out on
the town is this simple test: At the end of the evening, are the
people you were out with going to go home thinking, "That
TonyandDawn, what great people!" . . . or will they be thinking,
"That's the last time I want to go out with TonyandDawn"?

So what can you, as a couple, do to help ensure that everyone has a wonderful time? For starters, remember that everything Tony or Dawn does as an individual has a direct impact on Tonyand-Dawn. Big things *and* little things make a difference in how TonyandDawn impress the people they're with. When they both make a conscious effort to be inclusive, to ask other people their opinions, and to be interested in what they're doing, this gives others reason to enjoy being in TonyandDawn's company. On the other hand, acting superior or putting people down are instant mood-killers. No one has time for people who act like that. Finally, the person who crosses the line by flirting with or making passes at other people's partners is paving the way for long-term anger—and also risks permanently losing the couple as friends.

In the end, it's simple things, like the manners they use or don't use and their skill at carrying on a conversation, that affect how people see them as a couple. The way TonyandDawn settle the bill or chip in to share costs leaves an impression. Did they cross the line with their displays of affection? Did they ignore one couple all night long? Did they order the most expensive items on the menu? Were they dressed for a night out on the town? Did they spend all night gossiping and putting down absent friends? It all matters to each of them—because both Dawn and Tony know that any of these things can mean the difference between making and losing friends.

Who's Going With Us?

Dawn has called Marcy to discuss going out for dinner on Saturday night. There's a new restaurant in town that they've been wanting to try, and she's made a reservation for six people. Usually Jim and Jenny join the two of them and their husbands, but they're out of town this weekend. Marcy thinks of Bill and Sandy,

Check Before You Go

◆

Don't assume; ask.

Is that charity event you're going to black-tie or not? Who's driving?

If you have *any* question about *any* aspect of your evening, *check ahead of time*. Check (and double-check) when you'll be returning home, for the babysitter's sake. What time does the play start? How early should you arrive to get seated? What time are TonyandDawn supposed to pick RobandMarcy up?

To find out, all you need to do is pick up the telephone. If you're not sure, make the call and *be* sure, rather than risk making an embarrassing and annoying mistake.

---

a couple she recently met. "It might be nice to ask Billand-Sandy," she suggests to Dawn. "I know you haven't met them, but they're new in town and they seem really fun." So Marcy phones Sandy, and the date is set.

Old friends are great. There's a comfort level when you're with a couple you've known for years that can't be beat—but there's also much value to be gained from meeting new people and expanding your social set. Variety adds spice to your lives and to your relationship as a couple. And when you start to mix and match people from different groups, you're broadening your friends' horizons at the same time.

Deal-Breakers

Some social missteps are worse than others, and some can even leave your friends looking for new friends. The following behaviors are all certain to cast you in a very bad light—and, by extension, your partner.

A Superior Attitude and/or Putting Others Down

No matter what SidandLouise do, it never quite seems to be as good as what CherylandRob have done. If SidandLouise go to New York to see a play, CherylandRob will tell you about how they flew to London for the weekend so they could see the original cast perform. CherylandRob's kids are always the talk of the group . . . because they never let anyone else talk about their own kids' accomplishments. And no matter what work story others tell, CherylandRob always brokered the better deal, landed the bigger account, or told the boss a thing or two. Unfortunately, when one person starts upping the ante of accomplishments, others jump right in with their own stories. It dominates the conversation, and it's such a bore to watch the braggarts fluff their plumage all evening.

Of course, you wouldn't socialize with any couple that obnoxious—at least not for long. But the example helps remind us that the converse is true, as well: A couple that is modest about their doings is a couple that others enjoy being around. For example, we have good friends who never fail to ask about our kids—and their interest is absolutely genuine. They've watched our children grow up, and they've been involved and supportive every step of the way. Every time they ask what's new with our daughters, I know it's a sincere question, and I really appreciate them for it.

Similarly, while a little gossip is a tolerable thing among friends, the incessant derision of other people is a sure way to leave yourself out in the cold. Whenever Rachel runs into the woman she calls "The Mouth," she wants to run the other way. And that's unfortunate, because The Mouth's SO is really a very nice person. It would be a shame if Rachel and her partner had to give up on their friendship with the other couple because The Mouth makes it unpleasant to be with them.

ETIQUETTE IMPERATIVE

Leave Bickering at Home

◆

Fighting with your SO in public is a loser every time. There's nothing as unsettling to others as having to watch another couple get into a fight. People squirm, look the other way, try to change the subject—but in the end, a black cloud hangs over the evening. Guaranteed: Your companions will think twice before risking their good time by inviting you again.

The only thing worse is when the bickering couple tries to drag the people they're with into the argument: "Can you believe he said that? Marcy, am I crazy, or is he completely wrong? I know how you feel, because we've talked about this before . . ."

End of evening.

Equally annoying is the person who insists on criticizing his or her partner to friends. This seems to be a particular issue with women. Focus group participants have complained to us specifically about how uncomfortable they feel at being forced to listen to women rail about the shortcomings of their significant others.

Compliments, fun stories, reminiscences about past trips or experiences, trading ideas and observations—these are conversational topics that make you enjoyable to be with. Being a confidant can draw you closer to a friend—but when intimate talk turns into malicious gossip or catty remarks, it can drive a wedge into the friendship instead.

Finally, remember the cardinal rule of all good conversationalists: *Listen* to the people you're with. Being a thoughtful, engaged listener is the hallmark of a good friend and partner (see "The Art of Listening," page 36).

Flirting—and Beyond

There's flirting, and then there's behavior that goes beyond flirting. I'm talking here about the come-on—a direct, unmistakable pass. Once you've crossed this line, there's no going back. The relationship with that other couple will be forever altered. No matter what form it takes, coming on to someone else's SO is not funny, not cute, and not forgivable. Don't go there unless you're foolish enough to be willing to risk your own relationship along with your social and moral standing.

Even when you don't cross that line, flirting can be a confusing subject, because it often means different things to different people. In the Post survey conducted for my book *Essential Manners for Men*, however, a clear distinction did emerge between what is acceptable flirting behavior and what isn't.

First, it's important to define flirting, so that we're all working from the same page: If you enter into a conversation with another person *with the intent of starting a personal, one-on-one, evolving, and potentially* intimate *relationship with that person,* then you are flirting.

Now, with that definition in mind, as long as you *and* the other person are neither married nor in a committed relationship, flirting is a completely acceptable behavior. If, on the other hand, one of you is married or in a committed relationship, then your course of action is clear: Back off.

I've been known to joke and banter and have fun—perhaps even racy—conversations with women. All the same, I am *not* flirting. Why? Because I have no intention of having this banter proceed any further than a harmless conversation. The litmus test I have for my behavior and my conversations is simple: I never say or do anything with another person that I would feel embarrassed or awkward about if my wife heard or saw it.

The Flirting Litmus Test

Not sure if your actions are over the top? Never say or do anything with another person that you would be embarrassed or feel awkward about if your SO heard or saw it.

The key to flirting, in other words, is to know how far I can go in my behavior without my actions being upsetting to my spouse. These limits will vary with every couple, so be sure to know your partner's limits, and then respect them.

But why, you ask, shouldn't the burden be on your SO to accept *your* limits instead? One word: *Jealousy.* Jealousy is grounded in mistrust, and it breeds suspicion. These emotions and thoughts can only lead to misunderstandings and difficulties in your relationship. When your SO is secure in the knowledge that your behavior will always be within his or her comfort zone, mistrust and jealousy never have the chance to gain a foothold.

Failing to Observe the Niceties

While big things, like making a pass at a friend's SO, are obvious land mines, little things like manners matter as well. When small offenses accumulate, they can turn into deal-breakers all on their own. If you and your partner master the following essentials, you'll go a long way toward leaving a lasting positive impression on your friends.

Table Manners

Table manners rank above all other manners in importance. I think this has to do with the fact that eating involves performing an inherently unattractive activity—chopping up food on a plate, placing a bite-sized morsel on a utensil, attempting to lift it to our lips without spilling or dropping it, shoving the morsel into an open mouth, and then mashing it into a pulp and swallowing it—all while we're trying to carry on a pleasant conversation.

Table manners help us navigate this delicate task in two ways, by ensuring that (1) we are a good social participant during the meal and (2) we draw as little attention as possible to the actual act of eating.

Who Sits Where?

At a restaurant, when they're out with friends, couples usually split up, seating themselves in "boy-girl" fashion. After all, how often does Dawn get to focus on two wonderful men like Rob and Bill?

FLASHPOINT

Keeping Jealousy at Bay

◆

Jealousy indicates that you are not sure about your partner. Once it takes root, it is difficult to eradicate and can spring up again and again. If my wife had had jealousy on her mind when she found the note saying, "Was it good for you, too?" (see pages 22–23), the resulting difficulties would have been unpleasant, if not downright disastrous.

The antidote for jealousy is a very healthy, ongoing dose of communication, plus a willingness to modify any behaviors or words that are giving rise to the jealousy. Remember, though, that communication works only if it is grounded in sincerity. Assurances given with your fingers crossed behind your back aren't worth a plugged nickel.

Holding Chairs

Women like having their chairs held for them, or so they tell us in our *Post* surveys. In today's less-formal world, however, men often find themselves in a state of confusion: to hold or not to hold? The simple solution is to ask the woman who will be seated next to you what her preference is. As Tony approaches the table, he turns to Sandy, who will be sitting on his right, and asks, "Sandy, can I hold your chair for you?" Sandy can either say, "Why yes, thank you" or "No thanks, I've got it." Either way, the situation is resolved.

If Tony doesn't ask but instead simply reaches out to hold her chair, Sandy should accept the offer kindly, in the same spirit that Tony offered it—even if she prefers not to have her chair held for her. It's one of those small things that simply isn't worth making an issue over.

Making Conversation

Make a point of talking to the people seated on either side of you during the meal. Being a good conversationalist means focusing attention equally on each of your dining partners throughout the meal.

When conversation at the table is general, be a participant: Offer opinions and ask questions, but don't dominate the conversation or make leading comments that encourage others to engage in a debate or an argument with you. Also, be sure to tread carefully when discussing touchy subjects such as politics and religion. If you're not certain how your dinner companions feel about these topics, the safest path is to avoid them altogether. There's plenty else to talk about.

Staying Connected With Your SO

Small gestures—a discreet touch while passing your SO on the way to the restroom, a wink, a blown kiss, or a quick, private smile—are ways to let your SO that you are thinking of him or her, even when your partner is sitting across the table between your two good friends (see Chapter 4, "The Importance of Nonverbal Communication," page 43). If your sharp-eyed dinner companions spot the little signal, that's fine: It's all to the good for them to realize that you're a couple and that you like to express your feelings for each other.

More overt actions need to be undertaken with great care. For instance, it can be fun to play footsie with your SO beneath the table. Just make sure you have the correct foot.

The Basic Manners of Eating

The basic manners of eating fall into two categories: manners that make a difference and those that don't. Remember Emily Post's observation that it isn't which fork you use that matters— what matters is that you use a fork.

Eating customs that don't really matter but that you still may care to take note of include these:

+ Use silverware from the outside in.

+ Pass all food and condiments to the right, to avoid logjams and to ensure that everyone is served.

+ Pass the salt and pepper together.

+ Place your fork and knife side by side in the "four o'clock position" on your plate to indicate that you've finished eating.

+ It's okay to place your elbows on the table between courses.

- Eat with your right or left hand—it doesn't matter.

- Don't feel you have to eat everything on your plate.

- Place your napkin to the left of your place setting whenever you excuse yourself from the table.

- It's okay to *glance* at the handsome gentleman or knockout woman at the next table.

Table manners that really *do* matter include these:

- Avoid rude noises and emissions; excuse yourself from the table if you need to make any of these.

- Don't chew with your mouth open.

- Don't talk while your mouth is full of food.

- Don't take over-large bites of food.

- Don't eat as though you're afraid that someone is about to snatch your plate away.

- Don't dawdle over your food long after everyone else is finished.

- Be discreet about nose-blowing at the table: A gentle blow is permissible, but if you need to unleash a honking blast, excuse yourself to the restroom.

- Avoid talking too loudly or laughing uproariously.

- Never use your cell phone at the table to answer or make calls.

- Don't *stare* at the handsome gentleman or knockout woman at the next table.

When Someone Presses You for Your Opinion

◆

It happens to everybody: You're seated next to Pushy Patrick, and he has pressed you several times for your opinion on a hot political issue. You, on the other hand, don't want to offer your opinion on the matter, because you know the ensuing conversation is sure to escalate into an unpleasant exchange.

The solution is for you to turn to Pat and quietly say, "Pat, I know you have your opinion on the subject, and I have mine. But right now, this isn't something I want to talk about. Instead, I'd rather hear what you think about that new Steven Spielberg movie."

If Pat continues to press, turn to the people on your other side and start talking with them.

Bottom line: Never feel that you have to discuss a topic you're not comfortable talking about.

Excusing Yourself

Announcing, "Excuse me, I've got to go to the restroom" isn't necessary. Think about it—who really needs to know? A simple "Excuse me, I'll be right back" will do just fine if you have to leave for the restroom or make a cell phone call.

For the guys, whether to stand when a woman leaves or returns to the table can be a conundrum. Various Post surveys indicate that women like it when the men they're with (or at least the men on either side of them) stand on their departure or returns. That said, the question "Do I *have* to stand?" is hard to answer—and, worse, makes the gesture sound like a chore. A better approach is to make an attitudinal shift, so that the question becomes "What can I do tonight to make the women I'm with feel special?" Then, the solution becomes clear.

Similarly, if you're a woman and a man stands for you, don't think, *Ugh, I wish he wouldn't do that.* Instead, go for a more positive response, such as a simple "Thank you, John. That's so nice of you." You'll make his day, and yours as well. Just as important, you'll make yourself (and your SO) look good for your friends. Remember, it's all about having a good time.

Those Bodily Functions

---◆---

If Bill lets go with a major belch at the table one time, the others might laugh and excuse it. Do it two, three, or more times, though, and this first invitation might well be the last that BillandSandy receive from either TonyandDawn or RobandMarcy.

When people ask me what they should do about the urge to cough, sneeze, or pass gas during dinner, the question implies that there actually is an option other than excusing yourself to go to the restroom.

There isn't.

The bottom line: While certain bodily functions are unavoidable, they should be always done away from the table, where they won't have an unpleasant impact on the people you're with.

Paying the Check

If everyone in your group has ordered more or less the same amount of food and drink, then dividing up the check is easy: You simply split it evenly with the other couples, including a tip of fifteen to twenty percent.

If some of the people at the table order substantially more or less than the others, however, or if one or more couples indicate they'd rather get separate checks, then other arrangements are called for.

The Separate-Checks Option

If your group prefers separate checks for each couple and there aren't too many of you, ask for split checks at the start of the meal, *before* orders are taken, rather than waiting until the check is brought to the table.

Individual couples can choose this option as well. For example, Marcy and Rob have begun cutting back on big meals lately, and tonight they're enjoying soups and salads. Since the others are all ordering three-course meals, an even split of the tab is going to be unfair to Marcy and Rob. Again, to avoid any confusion later, they make their preference known right at the start of the meal. "We're on a soup and salad diet," Rob explains to the others, "so we're going to ask for a separate check. Make sense to you guys?"

The Fair-Split Option

I like the fair-split option best. After everyone orders, Bill and Tony realize that Rob and Marcy's share is substantially less than theirs. Bill says, "That all sounds great. Let's just make sure that when the bill comes, we split it fairly." Then, at the end of the meal, Tony reviews the check and says, "This looks correct. Rob and Marcy, why don't you chip in fifty dollars, plus ten dollars for a tip, and we'll split the rest of the bill between us?"

Thoughtful dinner companions will make a point of splitting the bill fairly with their lighter-ordering friends. But if they don't, it's also perfectly okay for Rob to suggest himself that he and Marcy put $50 (that is, the cost of their two meals, rounded up to a whole number) plus tip toward the dinner, and that the rest of the bill be split between BillandSandy and TonyandDawn.

Tipping

Notice that our dinner group is careful to get the tip straight between them. Each couple should add a tip of fifteen to twenty percent to whatever they are paying for their own meals, so that the total tip is commensurate to the total bill. To make sure they are all on the same page in terms of an appropriate tip, Bill says to Rob, "I make it out to be eighteen dollars for our portion of the tip. That's about twenty percent. Sound good to you?" Rob can then make his tip twenty percent as well, or $10 in his and Marcy's case.

When It's Time to Go

All good things have to come to an end, and the best way to end any evening out is on an upbeat note. The check's been paid, and the chairs have been held for the women (or not) as you're all getting up. Before you leave the table, do a quick check to make sure nothing has been left under a chair or on a seat. You could end up being the hero(ine) for a friend: "Marcy, I saw your compact on the floor by your seat. It must have fallen out of your pocketbook."

In cooler weather, your group will likely check everyone's coats. Each couple retrieves their own garments, leaving a $2 tip for the first coat and $1 for each additional coat. Tony helps Dawn on with hers, then quickly looks to make sure everyone else is all set. If Rob is in the restroom, Tony can help Marcy on with her coat as well.

Kiss, Kiss

Hello and good-bye greetings between friends often include a platonic kiss. Go for it. Between friends, particularly a man and a woman or two women, a light kiss on the cheek is polite and even expected. Just don't make it into something more than it is—a quick peck. At the same time, don't force anything. If you're more comfortable not kissing, either a handshake or a quick hug will do fine.

In the Car, Who Sits Where?

It's the little things that can often cause a moment of angst. Going to dinner, the plan was for TonyandDawn to pick Roband-Marcy up at their house and drive them to the restaurant. When TonyandDawn got to RobandMarcy's house, Dawn made a quick switch from the front passenger seat to the backseat so she could sit with Marcy on the way to the restaurant. It would have been perfectly acceptable for her to stay in the front seat, but she wanted to have a chance to talk to Marcy, so she moved.

On the way home, Marcy and Rob sit together in the backseat while Dawn joins Tony up front. If it turns out that Rob is six-foot-five, Dawn might say, "Rob, there's way more legroom up front. Please sit there. I don't mind sitting with Marcy." Rob can accept gracefully, or he can say, "Sitting in the back is no problem. Anyway, it gives me a chance to snuggle with Marcy on the way home." Everyone laughs—but more important, now everyone is comfortable with the seating arrangements.

15

AT A PARTY

WE LOVE IT WHEN A LITTLE SPICE GETS ADDED TO THE NORMAL
routine: The phone rings, and at the other end is an invitation to a
cocktail soiree, a Halloween costume party, or a dinner gathering at
the home of friends. Whatever kind of affair it is, the party will be
an event to anticipate beforehand and to remember long afterward.

I remember when I first heard about the Halloween bash.
Toby and Caroline were throwing a costume party for some
friends from the athletic club where we take yoga classes. My
wife, who had gotten to know a lot of the club members, was really excited about going.

I, on the other hand, still barely knew any of these people.
My "bah, humbug" persona immediately took firm hold: I glumly
imagined a party with a bunch of people I really didn't know,
other than when they were panting on an elliptical machine or

doing their ujayi breathing and holding mulabandha while sticking their rear ends in the air in Downward Dog. Not only that, but for only the second time in thirty-two years of marriage, we were going to dress up in costumes. My wife had already picked hers out: She was going as our black Labrador retriever. "What are you going to be?" she asked sweetly.

Finally, the big night arrived. Sporting a black suit, an oversized cigar, and a Bill Clinton mask, I entered the party with my "black Lab" on a leash (that part was interesting). The festivities were in full swing—the wine and beer were flowing, and tons of delectables were spread out on the table. The guests were sporting every costume imaginable: Caroline's mother was dressed as a witch, with incredible white makeup and luminescent spiders climbing through her hair. Hugh Hefner showed up with a Playmate on his arm. Some cross-dressing hula dancers, complete with coconut bras, burst in and stole the show. Then the music began playing, people started dancing, and suddenly it was 3:00 A.M. and we were the last ones saying good night.

We never made it to yoga class the next morning. That night, however, I learned some valuable lessons:

✦ It's always good to branch out and meet new people. We've since become good friends with many of the people we socialized with that night—both at the gym and outside of it.

✦ Stop bellyaching and get into the theme of the party. You'll enjoy yourself a lot more, and the other guests will enjoy having you there a lot more.

✦ When your SO really wants to do something, step up to the plate and *do it*. After all, you're together because you have similar interests. So, go for your SO's suggestion with an open mind and a willingness to enjoy. You'll both have a better time for it.

Before You Get There

When you get an invitation to a party, several issues of etiquette must be addressed before you even knock on your host's door: what to bring, what to wear, and when to arrive. Handle these issues correctly, and you'll get the evening off to a good start for everyone concerned.

"Can I Bring Something?"

Virtually every time we invite people over, the first thing they ask is: "What can we bring?" They're not asking because it's a potluck meal where everyone brings a dish (it isn't). Rather, they're making a genuine inquiry, born of their desire to help lighten the load for the hosts.

Although many people think they *have* to ask if they can bring something when invited to a dinner or a party, you may be surprised to know there is absolutely no requirement that you make such an offer. Unless it's a potluck gig, you and your SO should feel totally comfortable if all you do is accept the invitation, attend the party, and be great guests.

That said, if you want to make an offer, go ahead and do so. The worst that will happen is that you'll be politely turned down. For instance, when Roseann calls and invites Margeand-Tim to a party, Roseann responds to Marge's offer to bring something by saying, "Marge, thanks so much for asking, but this time I really just want you and Tim to come and enjoy the evening."

In this case, Marge should respect Roseann's wishes. If she simply can't help herself, then a nice bouquet of fresh-cut flowers, scented candles, a box of chocolates, or a bottle of wine is appropriate to have in hand when Roseann answers the door.

If Roseann instead replies, "How about a salad or a dessert?" then Marge is on the hook to deliver. Not bringing a side dish after saying they will would be a definite MargeandTim faux pas.

When to Arrive

"What's the single biggest etiquette mistake people make?" It's a question I get asked all the time. The answer is easy: Showing up late.

The issue is really pretty simple. Being late is essentially a mark of being disorganized, and this disorganization translates into a lack of respect for others. Lateness is especially frowned on in the business world, but it's just as disrespectful when you show up late for your tee time, the tennis court you reserved with another couple, or the dinner date you made with friends. Leaving people hanging is a bad way to kick off any activity.

"Ah," says my questioner, "but what about for a party?" Here's where the etiquette can get a little tricky. At most dinner parties and get-togethers with friends, you can be "fashionably" late and not ruffle any feathers. What's acceptable? Ten to fifteen minutes. Any later than that and you're making problems for your hosts, who are trying to plan an evening. If you are going to be later, call.

The bottom line: The host's wishes and expectations should guide your decision about when to arrive. Different hosts and different types of events may dictate either that an on-time arrival is called for, or that an arrival more than fifteen minutes late is reasonable.

Dressing

Casual is in, most of the time: At the backyard barbecue, the informal dinner party, the get-together with a few good friends, you'll do fine in sweaters and trousers, skirts or pants along with a nice top, and maybe a colorful scarf to dress it up.

If you're not sure, ask. It's perfectly fine for Marge to call Roseann, even on the day of the event: "Roseann, it sounds like you're having a major party. Are people dressing up, or is it casual?" Whenever we get a call like this, we simply tell the caller what I'm going to be wearing and what my wife will be wearing.

In addition to wearing inappropriate clothes, an even faster way to stop getting invitations is to show up smelly, dirty, or unkempt. Unless you're trying to make an avant-garde fashion statement, tangled hair, disheveled clothes, and (for men) two-day-old stubble is a look that should be saved for those nights when you're relaxing alone at home.

FLASHPOINT

How You Treat Your SO in Public

◆

Acting superior or in a demeaning way toward your SO (or anyone else, for that matter) will quickly put a chill in the air. Here, from our couples survey respondents, are a few reminders of how these actions can affect your partner, not to mention the other people you're with. (see "A Superior Attitude and/or Putting Others Down," page 170).

"He does not know how to disagree tactfully . . . so when a difference of opinion comes up, he likes to argue his point. That makes me uncomfortable, especially when there are extenuating circumstances that might explain the other person's point of view."

"He has a habit of talking over me when we're having a discussion."

"He puts me down and disagrees with statements I make."

"He talks a lot and doesn't listen to others enough."

"He ignores me in favor of other friends who are around. And he flirts with other women!"

"In party situations, he completely abandons me to be the life of the party."

Avoiding Conversational Blunders

MargeandTim ring the doorbell. RoseannandCharlie answer the door.

"Hello. Great to see you!"

"Thanks for coming. Oh, these flowers are beautiful. Come in, come in!"

Flowers are a wonderful gift, but they also fall into that category of gifts that require a little work—probably not what your hosts most want to deal with as they're greeting guests. So Marge offers to put them in a vase herself, while Tim handles their coats. "Just point me in the right direction," Marge says.

With flowers and coats taken care of, MargeandTim are ready to join the other guests. Now is the time when they can either shine or crash and burn, both in the eyes of the other couples present and in the eyes of each other. It tends to follow, by the way, that the more considerately and respectfully they treat each other, the better the impression they'll leave with the other guests.

"This Is Tim, My Husband"

At small, informal gatherings of friends, introductions are never a problem since everybody already knows everybody. At larger parties, however, there's a much greater chance that Marge or Tim won't know everyone else there. That means they each have a responsibility to make sure the other is properly introduced and included in any conversations that may have started.

While you'd think this would be a no-brainer, our couples survey respondents tell us that failing to introduce one's partner and include him or her in the conversation is all too common—and very frustrating:

> *"Because my husband can't remember names, he doesn't introduce me to people we may run into."*

"He forgets to introduce me when we run into some of his friends."

"My SO ignores me, especially when we're in a social situation and I am not introduced to an acquaintance of his."

"My partner leaves me standing alone without introductions."

"Sometimes he forgets to introduce me to people I don't know (as if I'm not there)."

You've Forgotten a Name

One of the most common reasons why people fail to introduce their spouses is because they've forgotten the name of the person they're talking with. So instead of making the introduction, they leave it up to the other person to introduce himself or herself to their spouse—hoping, in the meantime, that nobody notices their omission.

Bad idea.

In this situation, the simplest solution is to rely on the third principle of etiquette—honesty—and admit your temporary memory lapse. "Here's my wife, Ginny. I'd like to introduce you, but I'm really having a bad time with names this evening." Ginny, on hearing your predicament, can pick up the ball by extending her hand and saying, "Hi. I'm Ginny. Pleased to meet you."

Smart couples will actually take this a step further by setting up a little signal ahead of time: "If I rub my chin, it means I don't remember the person's name, so introduce yourself."

ETIQUETTE IMPERATIVE

Send a Signal

♦

Signals can be lifesavers. A touch on the back later that evening says, "I'm thinking it's time to go, okay?" A wink across the room lets your SO know, "I'm way over here with these people, but you're on my mind." Working out a few of these simple visual cues with your partner will help you avoid little glitches *and* make your partner feel special throughout the evening.

"What Do You Think, Dear?"

MargeandTim join a group in midconversation. Marge, being a little shy, remains an observer rather than a participant. She's starting to feel somewhat awkward just standing there. So Tim, seeing the problem, turns to her and says, "What do you think, Marge?" thus drawing her into the conversation.

It's a simple social gesture, and also one of the most important you will ever make. Whether the person being welcomed into the circle is your partner or another guest, one of the primary hallmarks of a good guest is a willingness to engage everybody in a conversation.

Not Interrupting

Imagine actually turning to an acquaintance and saying, "You know, Tom, your opinion really doesn't matter to me." If you did this, odds are it would cause irreparable damage to any relationship you hope to have with Tom. It's also doubtful that you and your partner would see much of Tom and his SO as a couple in the future.

When you interrupt someone, you are telling that other person that their opinion doesn't matter. Interrupting is one of the more frustrating and demeaning things people can do in conversation. Instead, be a good listener and wait until the other person's thought is finished before jumping in.

If you do interrupt someone, including your SO, immediately acknowledge your faux pas: "I'm sorry, dear—I interrupted you. What were you about to say?" Do it sincerely, and you're off the hook; do it insincerely, and you might as well not apologize at all.

Keep Your Dirty Laundry and Other Private Issues Private

Marge and Tim gain absolutely nothing by talking about their private issues at a party or other public event. No matter which partner does it, the other is going to feel embarrassed and frustrated at the lack of respect shown by this kind of talking out of school. Our couples survey respondents put it this way:

"He gets so involved in a conversation that he forgets what is private and what isn't. Sometimes he has boundary issues about how to share information with others, particularly about me."

"She tends to tell people more than they need to know about us, herself, and me. Sometimes I wonder if she thinks about what she's about to say, or if she does it for attention."

"Sometimes I believe she rather innocently moves the conversation into areas that I sense other people are uncomfortable talking about."

Leave Any Disagreements at the Door

Arguing with your SO is unpleasant enough in the privacy of your own home. Doing it in public makes both of you look bad and upsets the other people you're with as well.

Even worse is the fighting couple who tries to drag other people into the fray. It's a one-way ticket to not being included next time. (See Chapter 12, "When Reasonable People Disagree," page 139, for a full discussion of how to deal with disagreements.)

Booze

◆

Drinking causes all kinds of problems. In particular, from the point of view of etiquette, drinking can be a relationship killer. How many times have you seen someone who's had too much to drink, and then gone on to say and do things that embarrass themselves or, worse, really damage a relationship with their SO, a friend, or a business acquaintance?

Here are just a few of the many comments from our survey respondents about the effect that drinking too much can have on others. (My own observations follow in parentheses.)

"If he drinks a bit too much he can get obnoxious, and I do not like that." (If his partner doesn't like it, what are the other people thinking?)

"If he drinks too much . . . he thinks he's funny, but I don't find him funny. He makes what I consider to be stupid jokes with waitresses, and flirts in a way that I find annoying. I also think the other person thinks he's being annoying in an obnoxious drunken way. Fortunately, this very seldom happens." (Once is enough to ruin a relationship. There's no magic number of times that it's "okay" to be drunk.)

"My SO gets drunk, argumentative, and arrogant, and insists upon driving even though greatly impaired." (Being argumentative is a classic result of overindulging, and it's enough to be a problem on its own. Driving while intoxicated, however, is dangerous, criminal, and unacceptable.)

"If he's drinking, he acts stupid, and it's annoying." (Well said.)

"When we're out, he sometimes drinks a little too much. I am always the designated driver. I don't mind, but by the time we get into the car to come home, he passes out. Then, when he wakes up in the morning, he doesn't remember what has happened the night before." (Drinking enough to pass out? Not remembering what happened the night before? Not impressive to your partner or to your other companions—and dangerous to yourself.)

No Boys/Girls Allowed

◆

We never seem to outgrow that division of the genders we learned in grade school. How many gatherings have you been to where the women congregate in one area and the men in another? And yet, the best parties I've attended have always been ones where the two sexes mingle freely. Take the lead: Mix it up.

Being Good Partners

Being a good guest starts with being a good partner. Arriving at a party and then ditching your SO for the night is cause for trouble later on that night. Conversely, being a good partner reflects well on you and on the two of you as a couple. Suggestions for being proactively attentive to your partner include these:

✦ Take her coat, or she might take his coat, too. It's the offer that's important.

✦ Get her a drink.

✦ When he returns with the drink, welcome him into the conversation: "Thanks, Jim, that's so nice of you. You know Sally and Chuck. We were just talking about . . ."

✦ Bring him or her a plate of hors d'oeuvres to share.

✦ Make those little shows of affection like holding hands or winking or blowing a kiss (see Chapter 4, "The Importance of Nonverbal Communication," page 43) that say "I like being here with you" or "I love you."

✦ You don't have to stick together like glue for the entire evening. But do check in with your partner periodically to make sure she or he is having a good time. Excuse yourself from the group you are talking with by saying you want to get a refill, or simply say honestly, "I'll catch up with you in a little while. I want to ask Sue something."

✦ If he isn't having a great time, take note and offer to leave earlier rather than making him stay until the bitter end.

Being Good Guests

Being a good guest takes some effort. Why? Because it involves *thinking* about what you are doing, rather than just *doing* things without thinking. Following is a laundry list of things that you can do—both as an individual and as a couple—to make the evening brighter for everyone:

✦ **Take care not to damage or muddy property.** When you arrive, be especially careful to clean off your shoes and to shake off any excess rain or snow from yourselves and your belongings. During the party, don't prop your feet up on the coffee table, and do your best not to spill any food or drink on the furniture or the floor.

✦ **Be responsible.** If you break or spill something, tell the hosts right away. Help clean the spill up, and offer to replace the broken object or pay any cleaning expenses. Above all, be sure to apologize—and then be sure to write a note the next day as well.

✦ **Don't snoop.** Snooping is not okay. Don't go looking through the contents of your hosts' medicine cabinet, desk drawers, bedroom tables, or anything else.

✦ **Use coasters for drinks.** Glasses sweat and leave ring marks on furniture. If you don't see a coaster or at least a cocktail napkin to place under your drink, don't assume that means the hosts don't want you to use them. Don't use a book or magazine. Instead, ask for a coaster.

✦ **Follow your hosts' lead.** If they announce that dinner's ready, don't continue to sit in the living room for the next fifteen minutes finishing up your conversation. Get up casually, and go with the others into the dining area.

✦ **Practice good bathroom etiquette.** Put the toilet seat and lid back down after you're finished. Wash your hands, and then, after drying them with a hand towel, clean any water droplets around the sink and fold the towel neatly for the next person. If you've used up the last of the toilet paper, check under the sink for more. If there is none, let the hosts know right away.

P.S.—Before you begin using the toilet, check to make sure that there's sufficient toilet paper on the roll. It's always a bit embarrassing to have to shout out your dilemma for all at the party to hear.

✦ **Offer to help.** An offer of assistance is always a gracious gesture. Could you help refill people's glasses or pass some hors d'oeuvres? How about clearing plates away after a course, or perhaps pouring some coffee? If the hosts are close friends, you can even offer to help clean up afterward. If your offer is refused, don't press or forge ahead and do the task anyway. If it's accepted, then pitch in with a smile on your face.

✦ **Don't smoke in your hosts' home.** Ask your hosts where you can go to have a cigarette. When you're done, don't just toss the butt on the lawn. Dispose of it properly, and remember to check your shoes for mud or dirt or snow or grass clippings before going back into the house. Finally, think about carrying some breath mints in your pocket to freshen your breath before you start talking to the other guests again.

✦ **Stay home if you're unwell.** If you have the flu, a bad cold, or any other infectious ailment, it's unfair to your hosts and the other guests to expose them to your illness. Encourage your partner to go if he or she is healthy, but stay home yourself. Everyone will sympathize and feel ever so bad for you in your discomfort—and will also thank you many times over for your consideration. I know that whenever I see someone arrive at a party with red eyes, runny nose, and a box of tissue in their hands, all I want to do is run in the other direction.

Not Shaking Hands

✦

If you do have a cold and are at a party or in a situation where someone wants to shake hands, you can gently defer by saying, "Please excuse me for not shaking hands, but I have a cold and don't want to give it to you."

"Say Good Night, Gracie"

George Burns always signed off with that phrase at the end of *The George Burns and Gracie Allen Show.* Then his wife Gracie would laugh and smile and wave, and the show would be over.

All parties come to an end, too. And when they do, it's time to pack it in, thank the hosts, say good night to any lingerers, and head home. Try not to be one of the lingerers yourself.

The next day, remember to call and thank your hosts or, better yet, write a short thank-you note and send it in the mail.

Even among close friends, this is a wonderful way for guests to show their true appreciation for being invited to the party. When Leigh and Peter come to our house, Leigh always sends a thank-you note. And we always remember her kindness in doing so.

ETIQUETTE IMPERATIVE

Know When It's Time to Leave

◆

Don't overstay your welcome. I know, I know—you're having such a good time, you don't want the night to end. But your hosts do.

16

ENTERTAINING AT HOME

THE BASIC TENETS OF PUTTING ON A GOOD PARTY HAVEN'T really changed since the first edition of Emily Post's *Etiquette* was published in 1922. In that book, under the title "Detailed Directions for Dinner Giving" (pages 184–85), Emily wrote:

> *The requisites at every dinner, whether a great one of 200 covers, or a little one of six, are as follows:*
>
> **Guests.** *People who are congenial to one another. This is of first importance.*
>
> **Food.** *A suitable menu perfectly prepared and dished. (Hot food to be hot, and cold, cold.)*
>
> **Table furnishing.** *Faultlessly laundered linen, brilliantly polished silver, and all other table accessories suitable to the occasion and surroundings.*

Service. Expert dining-room servants and enough of them.

Drawing-room. Adequate in size to number of guests and inviting in arrangement.

A cordial and hospitable host.

A hostess of charm. Charm says everything—tact, sympathy, poise and perfect manners—always.

And though for all dinners these requisites are much the same, the necessity for perfection increases in proportion to the formality of the occasion.

Okay, I admit the part about the dining-room servants is passé in most homes today (though wouldn't it be nice?). The rest of Emily's advice, however, is still spot-on.

Planning Your Party

Successful parties and dinners don't just happen—they take planning and work. In fact, it's a pretty good bet that the more seamless the event, the more planning went into making it a success.

TadandGina, for instance, have decided they really want to have a party. It's reciprocation time: They have been to several people's homes and haven't yet had those people over themselves. Of course, they've been busy. But they also know that if they don't take a turn as host, eventually those same people are going to stop inviting them over. The entertainment street is definitely a two-way avenue.

Before they get down to planning the actual party, however, TadandGina have three basic questions to settle: When to have it, whom to invite, and what to serve.

When to Have It

It will do you no good to start planning for a party on a certain date, only to discover that your SO has to take a major exam, give a work presentation, or (as in my case) meet a deadline to hand in a 200-page manuscript the following day. Instead, pick a time when both people's schedules are relatively free, so that each partner can contribute to planning and hosting the party. (As it turned out, my manuscript was due to be finished on a Friday morning. On Saturday night we hosted a great dinner party, and one of the toasts was to the completion of the manuscript.)

Your children's schedules also need to be taken into account when picking your date. It's not fair to a kid (or her teammates) to suddenly learn that she's got to find her own ride to the soccer tournament or she won't be able to go.

Finally, there's always that issue of compromise to be dealt with (see "The Art of Compromise: It Won't Always Be Fifty-Fifty," page 137). My wife's brother-in-law has tickets to New York Jets games, and he regularly makes the five-hour drive from Vermont to the Meadowlands to watch them play. When it comes to choosing a date for a party in the fall, there is no confusion in that household: Game days take priority.

Whom to Invite

Emily Post placed great importance on creating the right mix of people for a party. "The proper selection of guests is the first essential in all entertaining," she wrote, "and the hostess who has a talent for assembling the right people has a great asset."

In the following passage in *Etiquette* (pages 185–86), Emily identifies some key issues that every host should consider when planning a party:

It is usually a mistake to invite great talkers together.
Brilliant men and women who love to talk want hearers, not
rivals. Very silent people should be sandwiched between good
talkers, or at least voluble talkers. Silly people should never
be put anywhere near learned ones, nor the dull near the
clever, unless the dull one is a young and pretty woman with
a talent for listening, and the clever, a man with an
admiration for beauty, and a love for talking.

Most people think two brilliant people should be put
together. Often they should, but with discretion. If both are
voluble or nervous or "temperamental," you may create a
situation like putting two operatic sopranos in the same part
and expecting them to sing together.

The endeavor of a hostess, when seating her table, is to
put those together who are likely to be interesting to each
other. Professor Bugge might bore you to tears, but Mrs.
Entomoid would probably delight in him; just as Mr.
Stocksan Bonds and Mrs. Rich would probably have interests
in common. Making a dinner list is a little like making a
Christmas list. You put down what they will (you hope) like,
not what you like. Those who are placed between congenial
neighbors remember your dinner as delightful — even though
both food and service were mediocre; but ask people out of
their own groups and seat them next to their pet aversions,
and wild horses could not drag them to your house again!

What to Serve

In choosing what to serve your guests and how to prepare it,
you've got lots of options. You can make everything yourself; you
can ask a few friends to "bring something" (typically a side dish,
such as dessert or salad); you can host a full-fledged potluck supper (in which every guest brings a dish); or you can have a caterer
handle all the food preparations for you.

Whatever option you choose, however, don't kid yourself: Even a potluck meal or a fully catered dinner requires a lot of planning and organization to be ready for when the food and guests arrive.

If you choose to do the lion's share of the food preparation yourself, here are three basic guidelines for putting on a first-rate spread *without* leaving yourself and your SO frazzled in the process.

Serve Food That's Easy to Prepare

First of all, serving food that's easy to prepare means serving food you've previously prepared for yourselves, so that you know what's involved in making it—and that it really is good. Second, it means choosing recipes that are relatively straightforward. You're asking for trouble if you select a menu that's going to take all day to prepare, knowing that you're both going to be busy at the same time trying to clean the house, check all your supplies, and get yourselves ready for the big event. The last dinner party we went to featured an absolutely delicious beef stew for the main course. All the hosts had to do was put all the ingredients in the pot and let it simmer for hours. They had plenty of time for other things that day, and still ended up serving a truly memorable dinner.

Serve Food That Can Be Prepared Ahead of Time

There's nothing more aggravating than inviting six or eight people over for dinner, only to find that you missed the entire cocktail hour because you were stuck in the kitchen doing meal prep. It's frustrating not just for you but also for the guests who've come to see you—only to discover that you're not there.

That being said, it's worth remembering that no rule in etiquette is inviolable. Recently, my wife and I threw a small dinner party for six people. Our kitchen has an island in the middle of the room, containing an old copper sink. Before the party began, we placed hors d'oeuvres on the island counter. Then, just before the guests arrived, I placed two bottles of Prosecco, a delicious Italian sparkling wine, into some fresh snow I'd taken from our yard and piled in the sink. The champagne glasses had been chilled in the snow as well, and then dried off. Our guests and I sat on stools or stood around the island, savoring the hors d'oeuvres and wine while my wife noodled around finishing the preparations for the meal. At one point, I actually asked if people would prefer to move into the living room, but everyone was happily enjoying the re-laxed informal atmosphere, the warmth of the kitchen, and the aromas of the veal stew simmering on the stove—and no one had any intention of going anywhere else.

Serve Food That Accommodates Your Guests' Idiosyncrasies Without Making You Crazy

There are all kinds of diets and special needs out there. When in doubt, the thoughtful host will ask guests whether they have any special dietary restrictions at the time the invitation is being ex-tended. That doesn't mean you have to change your whole menu to accommodate these special needs, however. The simplest rule of thumb when inviting people over is to decide on the menu *you* want to have, and then add on any additional items that might be needed to meet the requirements of various guests.

For example, TadandGina have invited JulieandChris to be a part of their dinner party, and Gina knows they're vegetarians. So in planning the menu, Gina makes sure she includes a cou-ple of tasty vegetable dishes and a salad. That way, everyone else can enjoy the duck à l'orange, and yet JulieandChris will be well fed, too.

Drinks

Every New Year's Eve my wife and I host a major, blowout celebra-tion. We even have a bonfire, weather permitting. Last year, the weather didn't permit—so we got to do the whole thing all over again on a perfect night in February. There's nothing like several dozen people standing around a large bonfire, with the sparks floating up into the crisp Vermont night air and a gentle snow falling at the same time. Everybody was oohing and ahhing appropriately, enjoying the warmth of the fire out in the middle of a hay field, when suddenly my sister-in-law let fly with a snowball. The game was on: Forty people began romping in the snow like kids, and everyone had a ball.

The interesting thing about that night's gathering is that some members of the group drink alcohol and some don't—and yet you couldn't have picked out the drinkers from the non-drinkers. Nobody drank to excess, and those who would be driving later were very careful not to overindulge. The result was that the evening was an unqualified hit.

The key is to decide just what you are willing to serve at a party and then be prepared with a good selection. Some people don't drink or serve alcohol at all, some serve beer and wine, and others will offer hard liquor as well. As the host, the choice is

ETIQUETTE IMPERATIVE

Be Responsible Because You *Are* Responsible

◆

Alcohol is a part of our society. So is your responsibility for the state of inebriation of any guest who leaves your house. It is not only your moral duty but (as recent court decisions have reaffirmed) your legal obligation to make sure each car leaving your party has a sober driver at the wheel. Nothing less is acceptable. Guests who have imbibed heavily and who don't have a designated driver may not like it when you insist on either calling a cab or driving them home yourself—but the next day they'll appreciate your concern for their safety and well-being.

yours. At my home we usually have available a variety of non-alcoholic drinks as well as beer and wine. I know one friend who enjoys Mount Gay Rum on a summer's eve, so we always have a bottle of that on hand, too.

The Devil Is in the Details

In Emily's day, households typically had a full staff to take on cooking the meal and cleaning the house. Today, my wife and I certainly don't have "staff"—and frankly, I don't know anyone who does. We've got to cope on our own. And that means . . .

Sharing the Load

When TadandGina first conceived of their party, they talked through every step of their preparations. First, they agreed, they would do the shopping together. Tad would talk with the people behind the meat counter about the filet he wanted. Would they trim it for him? They can do it in about five minutes, while if Tad does it at home, it's a forty-five-minute chore. For Tad, this is a no-brainer.

Meanwhile, Gina selects vegetables for the side dishes and cheese and crackers for an hors d'oeuvre. Julie is bringing some smoked salmon (Tad's favorite), and Heather has promised to show up with her outrageous spinach-and-bacon salad. Tina is supplying the apple crisp for dessert, so all the bases are covered.

Last but not least, Tad and Gina meet at the wine section and pick out several bottles of both red and white wine that they've had before and liked. After picking out some cut flowers to decorate the table, they're ready to check out and head home.

By shopping for food together, they not only cut down on the time it takes but also underscore the fact that this dinner party is a joint effort. This is especially important to Gina, who is frankly concerned that once the party begins, she doesn't end up waiting on the guests *and* Tad.

Back at home, Tad prepares the filet, slathering it in minced garlic and fresh ground pepper. It's now all ready to start cooking. When the proper time comes, Tad will simply pop it into a hot oven, at 425 degrees. Ten minutes later, he'll reduce the heat to 350; then, after another twenty minutes, he'll start testing the filet. An instant meat thermometer is one of the best ways to test for when a cut of meat is done. An average temperature of 125 degrees in the center is perfect for a medium-rare steak. The thick parts will be a little rarer, and the thinner parts will be a little bit better done—something for everyone. (Some people, of course, prefer their steak well done, and others like it extra-rare. Check the markings on the meat thermometer or a cookbook for the temperature that matches your own preference and then adjust the cooking time accordingly.)

Meanwhile, Gina trims the asparagus and prepares the baby carrots the way her mother taught her. Sweet potatoes can be cooked quickly in the microwave, and add a splash of rich orange color to the plate. Finally, a hollandaise sauce pulls it all together. Add Heather's spinach-and-bacon salad, and this is turning into a meal fit for royalty.

The best part of the meal is that the prep time takes about an hour. The vegetables, sweet potatoes, and hollandaise can be kept ready on the hot tray. The only task remaining is for Tad to put the filet in the oven about 40 minutes before meal time. Together, TadandGina will only have to make a couple of trips to the kitchen, so they'll be free to enjoy the cocktail hour with their guests. Plus, by prearranging who will prepare which parts of the meal and then jointly getting the house spic and span, they've also left themselves some time to relax before getting ready for the party.

Being the Gracious Host

TadandGina have already set a course that bodes well for their party that night. They've planned a meal that's delicious and easy to execute, and they've done a good job of sharing the load. Now, stir in some old-fashioned courtesy and a dash of sincere interest in making their guests feel comfortable and welcome, and the recipe for a successful party is complete. By adhering to the following checklist, TadandGina can help ensure that every one of their guests feels that inner glow that says: "Hey, I'm having a terrific time."

✦ **Be on time.** This key etiquette item doesn't apply just to guests. If Tad is in the shower and Gina is still drying her hair when the first guests arrive, the situation is going to be awkward. Guests may arrive fashionably late, but Tad and Gina need to be ready *before* the appointed time.

✦ **Greet your guests.** Don't let other guests answer the door. TadandGina, or at least one of them, should open the door for each arrival. They should be prompt about it, too. The worst thing any host can do is leave guests standing outside the door, wondering if they should just come in.

✦ **Be consistent.** When Gina made her calls inviting people, she made it clear that the dress was casual. For Gina to then open the door wearing a long skirt, a beautiful silk blouse, and her finest jewelry would not be appropriate, for the simple reason that it would embarrass her guests.

✦ **Be the majordomo.** During the party, Tad and Gina should circulate among the guests, checking that people have everything they want. They should also look for anyone who is alone, who appears to be trapped in a conversation they want out of, or who doesn't look like they're having a great time.

✦ **Be a leader.** Their guests will look to TadandGina for signals. At meal time, once a few people have been served, Gina announces, "Please start eating as soon as you get to the table. I read somewhere that Emily Post herself always wanted guests to start eating right away, so their food wouldn't get cold."

✦ **Check out the facilities.** Before the party begins, make sure toilet paper is in good supply, and that you've stocked up on any other incidentals guests might need. A small can of air freshener will save guests needless embarrassment. Fresh soap in the soap dish and clean towels to dry hands are a must. During the party, an occasional quick visit to the bathroom will ensure that plenty of supplies are available and that the room is still fresh and clean.

Appreciating Your Partner's Efforts

It's one thing to work hard preparing the meal and orchestrating the gathering. But it's the knowledge that your efforts are truly appreciated that makes the hard work all worthwhile. When it comes to putting on a party, Tad realizes he's a lieutenant to Gina's general. But he also genuinely appreciates his wife and everything she's done. So at the start of the meal, when everyone has been seated and all glasses are filled, he offers a toast:

"I just want to first thank Gina for making tonight happen. Gina, you are the best—and tonight you showed all our friends why. So, here's a heartfelt 'thank you' and a warm toast to the love of my life, Gina."

As I said at the start of this chapter, entertaining is a two-way street.

Entertaining Houseguests

When JulieandChris got back home from the party at Tadand-Gina's, they had a message on their answering machine. It was from their friends JayandDebbie: "We're so excited that you invited us to visit. Next weekend is great for us, if that still works for you. Give us a call tomorrow, and we'll firm up the plans."

JulieandChris realize they'll be sharing their small home with JayandDebbie for the weekend, and want everyone to have a great time, as do JayandDebbie. To make this happen, both couples must be prepared to work at being good hosts and guests, respectively, and must also be ready and willing to compromise for the enjoyment of all.

Five Steps to Being a Great Host to Houseguests

The key to hosting a successful overnight stay is to go in with the expectation that everyone is going to have a good time. Here are five suggestions for making sure this expectation is fulfilled, and that your guests' visit is enjoyable and glitch-free:

1. **Set clear start and end times for the visit.** When Chris calls their friends back the next day, he confirms with Jay that JayandDebbie will arrive on Friday in the late afternoon and will leave on Sunday afternoon. Chris then tells Jay he'll e-mail directions to their house later that day.

2. **Make sure your house is clean.** JulieandChris spend Thursday evening policing the house to make sure it is spruced up. Besides vacuuming and picking up any stray magazines or newspapers, they also put away the stack of bills and personal papers on the desk in the study and make sure the photo of the four of them at JulieandChris's wedding is displayed.

3. **Have your guests' room ready for them.** Well before JayandDebbie arrive, the bed in the guest bedroom will have been carefully made up with fresh sheets. There is an extra blanket at the foot of the bed, a reading light and alarm clock in working order, a box of tissues on the nightstand, and a wastebasket in easy reach. In the bathroom their guests will be using, Chris makes sure there are towels, soap, water glasses, plenty of toilet paper, and a box of tissues. He's also added a few extra touches: a vase of flowers, a couple of magazines, and bottles of shampoo and bath lotion.

   When JayandDebbie arrive, JulieandChris show them to their room. Then, after they are settled, Julie gives them a quick walking tour through the house so they'll know where everything is. In the kitchen, Julie points out the foods that JayandDebbie can help themselves to, as well as any foods that are being saved for a meal during the weekend — the blueberries in the fridge, for example, will be going into the pancakes on Sunday morning.

4. **Make plans carefully.** Chris knows that having houseguests means he'll probably have to bow out of his regular Saturday golf foursome. For this weekend, he can either make a time for Jay and himself to play a round or arrange

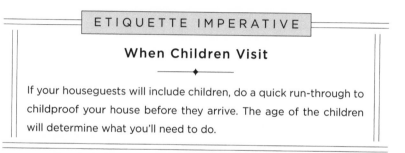

ETIQUETTE IMPERATIVE

When Children Visit

If your houseguests will include children, do a quick run-through to childproof your house before they arrive. The age of the children will determine what you'll need to do.

for the two couples to play. Before he sets anything up, however, he checks with JayandDebbie first to make sure they want to spend an afternoon playing golf. As a backup, JulieandChris scan the paper for local events and things to do. Options in hand, they then call JayandDebbie to say that they can get four tickets for that Saturday's double-A baseball game if they're interested. If they aren't, JulieandChris then offer some other ideas. Alternatively, JulieandChris may have already bought the tickets. In this case, their invitation will include the expectation that JayandDebbie are coming in order to go to the game.

5. **Leave some downtime.** No matter how many wonderful things are going on that weekend, don't schedule the visit down to the very last minute. JayandDebbie may be known for their high energy level, but even they need a little time to relax and catch their breath.

Five Steps to Being a Great Houseguest

In *Etiquette* (pages 435–36), Emily Post wrote about the difference between the good houseguest and the houseguest no one invites back:

> *The guest no one invites a second time is the one who runs a car to its detriment, and a horse to a lather; who leaves a borrowed tennis racquet out in the rain; who "dog ears" the books, leaves a cigarette on the edge of a table and burns a trench in its edge, who uses towels for boot rags, who stands a wet glass on polished wood, who tracks muddy shoes into the house, and leaves his room looking as though it had been through a cyclone. Nor are men the only offenders. Young women have been known to commit every one of these offenses and the additional one of bringing a pet dog that was not house trained.*

> *Besides these actually destructive shortcomings, there are evidences of bad upbringing in many modern youths whose lack of consideration is scarcely less annoying. Those who are late for every meal; cheeky others who invite friends of their own to meals without the manners or the decency to ask their hostess' permission; who help themselves to a car and go off and don't come back for meals at all; and who write no letters afterwards, nor even take the trouble to go up and "speak" to a former hostess when they see her again.*

With that warning in mind, here are five tips that will help make JayandDebbie the kind of houseguests who get invited back time and again:

1. **Know when to arrive and when to leave.** Suppose Chris said to Jay, "It's great that you're going to take us up on our invitation. Stay as long as you want. We love having you guys." Jay should then clearly reply, "We're excited to be coming. We want you to know that we'll be arriving on Friday in the late afternoon, and that we have to head back to Boston by midafternoon on Sunday. We're really looking forward to a trip to the country."

2. **Bring a house present.** Jay looks at Debbie after he gets off the phone. "Think we need to bring something? It's just two nights." Debbie just smiles. She knows (and he does, too, really) that they'll start looking for the perfect house gift that week.

3. **Go with the flow.** If JulieandChris already have tickets to the double-A ball game, JayandDebbie will go cheerfully, even if baseball isn't their favorite spectator sport.

4. **Make an effort to contribute.** JayandDebbie are there to visit and spend time with JulieandChris. If they wanted to be

House Present Suggestions

When staying at someone's house, a thoughtful gift for your hosts underscores your appreciation for their hospitality, and also serves as a pleasant reminder of your gratitude even after your visit is ended. The best approach is to arrive with a gift, but it's also acceptable to buy something during your stay. A third option is to send your hosts a gift after you've returned home.

Appropriate gifts for an overnight visit include

+ A bottle of good wine
+ A new, best-selling book
+ Scented candles
+ Golf balls for the avid golfer

Longer stays warrant a more substantial gift, such as

+ Hand towels for the powder room or monogrammed beach towels for sunning
+ A bottle of liqueur or cognac you know your host likes
+ An ice cream maker and toppings for sundaes
+ A lobster pot and the utensils for eating lobsters; then go to the market with your host when you arrive, and buy fresh lobsters and other fixings for a home clambake
+ A gift basket filled with specialty food products or the fixings for a gourmet picnic
+ A coffee table book about a favorite subject of your hosts'
+ Cocktail napkins monogrammed with your hosts' initials

A Third Thank-You?

◆

When JayandDebbie return home Sunday evening, a quick call to JulieandChris to let their hosts know they've gotten back safely is also the perfect opportunity to express their thanks once again for a wonderful weekend.

waited on hand and foot, then they should have planned a weekend trip to a resort instead. As guests at JulieandChris's house, they'll need to step up to the plate and do their part. They start by keeping their room and the bathroom neat and clean. They also help with dishes. They also ask if they can help prepare the meal. And if an errand needs doing, they are quick to volunteer.

5. **Thank your hosts twice.** As they get ready to leave on Sunday, JayandDebbie thank JulieandChris for their hospitality, and enthuse about the wonderful weekend they've had. Handshakes, hugs, and kisses are exchanged all around, and then they make their exit on time. The next day, Jay and/or Debbie write their hosts a thank-you note. They handwrite the note, put it in an envelope, and mail it. This is *not* the time for an e-mail. Taking the time to handwrite and mail the note makes it special. It will stand out from the bills and catalogs in JulieandChris's mailbox, and will be a perfect reminder of the great weekend they all shared.

17

EXTENDED-FAMILY DYNAMICS

ONE PIECE OF ADVICE THAT I HEARD IN A RECENT FOCUS GROUP keeps sticking with me. We were in the middle of a discussion about coping with extended family, and this person said: "When you're in a relationship, your SO's family comes with the package. Withdrawing doesn't work."

That is the heart of the matter. When you marry or move in with someone, you don't just gain a life partner—you also get your SO's mother, father, and siblings. And, of course, you bring your parents and siblings into the mix as well. One of the biggest challenges in being part of a couple is to learn to manage your relationships with both sets of parents and siblings in a way that doesn't collide with the person you want to make a life with.

It's kind of like merging onto a highway at rush hour: When you're all going in the same direction at more or less the same speed, the traffic flows relatively smoothly. Once in a while someone may cut someone off or get a little too close and put everyone's nerves on edge for a few minutes, but things quickly settle back down, and the traffic resumes moving normally.

When a car disrupts the pattern, however, by stopping cold on the on-ramp or, worse yet, pulling onto the highway headed in the wrong direction, then all hell breaks loose. Horns blare and tempers fray as drivers scramble to take evasive action. At best, traffic becomes hopelessly snarled; and if the driver's error is bad enough, serious damage can occur—the kind of damage that may not be able to be undone.

Relations with in-laws work in much the same way. When everyone is pulling in the same direction, things go fairly smoothly. But if a father- or mother-in-law or a sister- or brother-in-law won't accept their family member's partner, or when that partner refuses to be involved with his or her SO's family, trouble starts brewing.

Smooth Merge, or Crash and Burn?

When it comes to dealing with your SO's family, everyone's situation is unique. In reviewing the comments of our couples survey respondents, however, we saw several themes pop up constantly: the value of communication, the need for all parties to be willing to work at interfamily relationships, and the importance of recognizing and learning to tolerate different family cultures.

Our respondents also noted another important truth: *circumstances change over time.* Relationships with extended family members that start out well can gradually deteriorate if they're not tended to; and, conversely, a difficult or strained situation can, in time, grow into a positive (or at least tolerable) relationship. Depending on how they're handled, the arrival of children, the sharing of family experiences, and growing familiarity with each other all have the potential either to strengthen your bonds with your SO's family—or tug them apart.

Communicate

Clearly, when people make a genuine commitment to communicate, they have a much, much better chance of making their relationship with extended family work. Our survey respondents told us that whenever the lines of communication remained open, even the frostiest relationships between in-laws thawed over time.

Our survey also found that the biggest impediment to communication occurred when one or both parents refused to accept their child's choice of partner. If a parent's (or other family member's) starting position vis-à-vis the partner is one of outright antagonism, it will take a huge amount of time and commitment on the couple's part to change that position.

Our respondents also noted the vital importance of communication *between the partners themselves*, in order to maintain perspective on any extended-family issues they might be facing:

> *"We allow each other to talk freely about our feelings toward our parents, so that our relationship stays primary."*

> *"Both my husband and I are aware of the personality differences between my mom and him, and between myself and his dad. We're able to have nonhurtful discussions about it, as well, but we're each careful of what we say about the other's parent."*

Without these kinds of discussions, difficulties with each other's families can all too easily morph into conflict between you and your SO.

Willingness to Work at the Relationship Over Time

The couples responding to our survey indicated repeatedly that even when the mix between their SO and their parents was like oil and water at first, the combination of time plus consistent effort virtually always made a difference. But this consistent effort can arise only out of a sincere desire to commit to the process over the long haul. One good phone call or evening together isn't going to miraculously change people's attitudes and feelings. Over time, however, these and other constructive steps begin having a cumulative effect.

"My husband's parents weren't accepting of me for a long time. Even after we were engaged, they only introduced me to people as his 'friend.' We all knew that our troubled relationship was very difficult for my husband, and we've worked very hard on getting along for his sake."

"Getting along with each other's family is definitely an issue in our relationship. We've resorted to my SO doing most of the communication with his family, and me with mine. That seems to have taken some of the pressure off."

"We live close to my family. They didn't like my SO from the beginning, because he's a 'hippie' type. But they've come to accept him after fourteen years."

Different Backgrounds, Different Personalities

It stands to reason that different cultural backgrounds, political orientations, and socioeconomic parameters can lead to clashes between partners and families. If two partners come from very different backgrounds, or if the personality of one partner differs significantly from the personalities of his or her in-laws, the couple will invariably face added stresses in building relationships between families. Our survey respondents were quick to point out how difficult it could be to get along under these circumstances:

"I don't care much for his family. They are very different from mine."

"I don't really like his family. He comes from a much bigger and louder family than I do, and so I need a nap after spending time with them."

"My family isn't very social and didn't accept his outgoing family at all."

"Different cultural backgrounds sometimes clash. He has a large family, and they demand a lot of attention."

"He likes my family more than I like his; this stems from the different values that our families have."

"His family have some traditional views that he and I both don't always agree with."

"My SO's family is of a different religion, and would not come to our church ceremony because of this. I have never met them, and I hope it stays that way."

"We come from very different family backgrounds. This caused a lot of stress in the first year of our marriage, but things have now calmed down."

For all the difficulties that sharp differences can cause, however, our respondents also told us that such relationships *can* be made to work, if parents and couples are willing to be tolerant and plug away at the relationships.

"She is Jewish and I'm Catholic, but each family accepts the daughter-/son-in-law, and encourages them to participate in all family activities, including religious and nonreligious functions. There is a great deal of respect between families for their history, traditions, and culture."

The Holiday Season

My wife and I were at a New Year's Day brunch when the subject of how families decide where to spend the holidays came up.

"That's easy," said Stan. "I just say, 'Yes, dear.' "

Everyone laughed in recognition.

Couples develop a wide variety of ways of handling both where they go for the holidays, and how they make the decision on where to go. There simply is no set rule for working this out, other than perhaps the most important rule of all: communicate.

It is essential that you communicate not only with your SO about your holiday plans, but also with your family and her family as early as possible. That way, you avoid the stress of those last-minute guilt trips that one parent or another can so easily hit you and your SO with.

One survey respondent reports that she and her husband will go so far as to decide their holiday itinerary a year in advance: "We discuss it well before the holiday, and decide together how we'll manage it. Sometimes, we decide during this year's holiday what we'll be doing next year, so there's no question and nothing to stress over."

On reading through hundreds of comments from people about how they handle the holidays, it's clear that the most successful couples employ several key strategies:

✦ **The bottom line is, you and your SO—and your
   relationship—come first.** "We just try and be fair to each
   other's families and to our own. We tell our parents in a
   matter-of-fact way that this is what we would like to do for
   the holidays, keeping in mind and respecting what each set
   of parents would like to see happen at the holidays." (This
   couple understands the importance of both respecting their
   own desires and also being cognizant of the hopes of their

parents and other family. Still, the decision they ultimately make is theirs, rather than someone else's.)

✦ **Be flexible.** "Both of our families are out of state, so we need to travel for every holiday. We both come from big families, but traveling has *never* been an issue. Certain holidays mean more to us and our respective families than others. We just go where we feel we need to be, when we need to be there. And since the arrival of our kids, our own family comes first." (This couple understands that circumstances change for themselves and for others. They consider the current situation when making their plans — and, adhering to the first strategy, they also focus on the needs of their own family first.)

✦ **Make no guarantees.** "We're very fortunate that the holidays have never been an issue for us. It all just fell into place. Now that we're married, I make no guarantees on how it will work each year. I feel confident that this, too, will fall into place." (Change is the only constant — and you and your SO need to recognize that as circumstances change, your decisions on what to do will change as well.)

✦ **Know what's most important to each party:**

  ✧ "We spend Thanksgiving with my SO's family every other year. Then we spend Christmas Eve with my family, and Christmas Day with his." (Many survey respondents reported that they alternate the holidays in just this way. Touching every base is a good way to avoid ruffling any feathers, but be mindful of the effect it can have on you and your SO. As one respondent noted, "Yes, we divide the time between both families — and we just end up exhausted.")

⊹ "Thanksgiving is a bigger deal with his family than it is
with mine, so he has dibs on this holiday. Christmas
Eve is a bigger deal in my family, so I have dibs on that
one. New Year's is just for the two of us." (Sharing,
understanding, and remaining committed to themselves
as well—this couple has it figured out.)

Staying Home

Many couples also find refuge by simply staying put for the holidays. The switchover from visiting relatives to hunkering down at
home mostly seems to take place around the Christmas holiday
season, especially after children enter the picture. When we were
first married, my wife and I traveled to New Jersey every other
year to spend Christmas with her family. Once we had children,
her mother told us she didn't expect us to make the trip. "Children should enjoy Christmas in their home, with their own traditions, just like you celebrated Christmas at home when you
were growing up," she explained unselfishly to my wife. Those
traditions have grown and have become very important to our
daughters. At the same time, we've found other holidays to share
with my wife's family. Thanksgiving is now a time when my in-
laws traditionally come up to Vermont to visit us; and for Easter,
we travel to visit them in New Jersey.

Separate Holidays

Almost without exception, the answers that couples gave to the
various questions in our couples survey did not vary significantly
by age or marital status. One place where marital status clearly
does have a large effect, however, is on what people choose to do
for the holidays—specifically, whether they decide to spend the
holiday season apart from their SO. Only one married couple in
the entire survey reported celebrating the holidays in separate

Holiday Vacations

———◆———

Some couples avoid the whole question of where to go for the holidays by taking off on an end-of-the-year vacation. "We always go on vacation somewhere warm, and never with extended family," wrote one couple. "We see them enough throughout the year that we don't need to dedicate the holidays to them. That is our time, and we always get away."

However, while heading off on vacation certainly clarifies your holiday itinerary by neatly sidestepping the issue of who to visit, be aware that family traditions may suffer as a result. When I was growing up, my family always went on a winter vacation to a ski area, but we didn't leave until the day after Christmas. One year, when my brother and his wife were in Liberia with the Peace Corps, my parents took me, my sister, and a friend of my brother's to visit them. We ended up spending Christmas in some very rustic guest accommodations in Liberia. As unforgettable as the trip was, I missed Christmas without the tree and the familiar traditions and all the family we'd normally see. I haven't been on a holiday vacation since.

Then again, some couples go a step further by simply deciding to ignore the holidays altogether. When I was a guest on a radio show recently, one caller began his question by saying, "I choose not to celebrate the holidays." Not a choice I could ever see myself making—but if it works for him . . .

places. Meanwhile, about one out every six couples who were in committed relationships but not yet married still went their separate ways for the holidays. Here are a few of their comments:

> *"I go with my family and he either joins me or not, and it is fine."* (Age: both partners in their fifties. Status: committed relationship.)

"I usually spend one holiday with my family and spend Christmas with his. He always spends holidays with his family, and not mine." (Age: both partners in their late twenties. Status: living together.)

"We usually end up separated because neither of us wants to spend it away from our family." (Age: both partners in their early twenties. Status: committed relationship.)

Balancing Family Traditions

In addition to the question of where to spend the holidays, the issue of honoring various family traditions is also important. Prior to hooking up with your SO, you followed certain holiday traditions in your family of origin, and so did your partner. As a couple, you need to decide which of your family's traditions you'll continue to uphold and which ones you're going to let go of because they conflict with your SO's traditions. For example, my wife's parents decorated the Christmas tree only after the children had gone to bed on Christmas Eve. Looking back now, I don't quite know how they found time to decorate the tree, what with seven children's presents to wrap and stockings to fill, plus a house to clean up. But somehow they did—and every Christmas morning, the kids would come downstairs and discover the miraculous transformation.

Great for the myth; tough on the parents. We chose to drop that particular tradition and decorate the tree prior to Christmas Eve—not a tough decision.

In my family, we children used to wake to find our filled stockings at the foot of our beds. Needless to say, we ripped into them right away. In my wife's family, on the other hand, the kids would find their stockings hanging on the fireplace mantel, and would then open them all together. In our own home, we opted

ETIQUETTE IMPERATIVE

Limits, Limits, Limits

My parents are coming to visit?!

Or . . .

We're going to your parents' for the weekend?!

Two inviolable rules govern these visits. First and foremost: Before either of you even raises the possibility with your family, discuss any plans you're thinking of making with your SO.

Second: Establish a start and end date for the visit *at the time the invitation is extended.* No open-ended visits. Repeat: No open-ended visits. No exceptions, and no excuses for not setting the parameters clearly.

(For other guidelines to hosting or being a houseguest, see "Entertaining Houseguests," page 210).

to hang the stockings on the mantel and to enjoy the experience of opening them with our kids on Christmas morning.

The key in both instances was that we respected both sets of traditions in making our choices. My wife and I each had certain customs that were important to us, and so we worked together to figure out how best to blend these customs, and how to develop our own family traditions as well—traditions that we hope our daughters will bring to their families in the future.

18

WORK—THE OTHER SIGNIFICANT OTHER

EVERY WEEKDAY MORNING, JOE RISES AT 5:30, PADS QUIETLY into the bathroom, showers, shaves, and gets dressed for work. He's always careful to keep the noise down, so he doesn't wake Nancy or the baby. And then he's gone.

Invariably, he gets home at 6:00 in the evening. He's usually a little tired, but the sight of their tiny son is more than enough to perk him right up. While Nancy gets dinner ready, Joe plays with little Joe, bathes him, and gets him ready for bed. At 7:30, Joe and Nancy sit down to a nice quiet dinner together. They talk about the baby, about the house, about their plans for the weekend. In fact, they talk about every imaginable subject—except for Joe's work.

Nancy knows her husband works as an engineer for a construction company. But she has never been to the office where he usually works, or to any of the construction sites, either. Before the baby was born, when she had her own career, they never had the time or the opportunity to visit each other's workplace. Now, with a child to take care of and a part-time job of her own, she realizes she still has no concept of what his life is like on the job.

It's a mystery—and Joe really doesn't want to talk about it. "When I'm home, I don't want to think about my work or talk about it," he explains to Nancy. "I just want to be with you and little Joe."

One of the greatest dangers to a relationship is an active imagination. Jealousy is all about grappling with the unknown, and using imagination to fill in the blanks. It's a small step from Nancy wondering what Joe is doing all day to becoming jealous of the time he spends at work. I'm not talking here about jealousy over who he might be spending time with at work or what he's doing after work ends. I'm talking about a growing jealousy aimed at his job itself.

Many people never see their partner's place of business. As a result, they have no concept of what their SO's work is like or what kind of pressures he or she faces. The best way for each of you to demystify this place (that gets more face time than your partner does) is for you to talk about it:

- ✦ Describe the people.

- ✦ Vent a little about your frustrations.

- ✦ Boast a little about your successes.

- ✦ Confide in your SO about your hopes for promotions, bonuses, and increased responsibilities.

- ✦ Invite your partner to visit you at work so that he or she can visualize you during the day.

In short, break through the mystery and make your work world real for your partner.

Working Together

On December 1, 1984, I opened the doors to my new advertising agency in the Maltex Building, an old factory that had been converted into a place for incubating small businesses. I had about 1,500 square feet of open space, huge windows with magnificent views of Lake Champlain, beautiful hardwood floors, a bunch of boxes that arrived just that morning, and a typesetting machine.

By the first of February, the office had gained a reception area, defined by waist-high walls and a growing mountain of mail that was piling up because I was too busy focusing on getting my work done. Then my wife stopped by for the first time. She eyed the vast pile of mail—including bills, checks, and who knows what else. "I know you," she said. "You're never going to get this stuff done."

Of course she knew me: We were married. And she also knew what my desk looked like at home. So she went through the mail, sorting bills, depositing checks, and generally getting things organized. And just like that, she had a new job. We've worked together ever since, and continue to do so today in our jobs at the Emily Post Institute.

I count myself twice lucky on that score. First, I get to spend time with my wife during the day. Second, somehow we've managed to survive working together without going at each other's throats. I can't tell you how many people have told me that they could never work with their significant others. I understand exactly what they are saying. I also feel sorry for them. Being able to be together both at work and at home has been great for us. In the process, we've also learned a few things about mixing work and relationships:

✦ When you're at work, you have specific roles to play, and both of you must be comfortable accepting each other in these work roles. At our ad agency, I was the head designer and salesperson, while my wife was the chief financial officer. For the sake of the other employees and our business, those were our roles when we were on the job — not husband and wife.

+ Once you leave the office, your roles switch to being partners in a relationship, regardless of what your roles might be at work. You have to be comfortable jumping back and forth from your work roles to your partner roles, and be willing to make that transition as soon as you walk in or out the door of your workplace.

+ When you leave work, leave work behind. That guideline sounds good in principle, and we try hard to adhere to it. But the reality is that we talk shop at home, too. The benefit is that we get things sorted out and figured out, so that the next day we're already moving forward. The danger is that work can all too easily encroach on and take over our personal lives together. As long as the business is our business, a *little* of the work-at-home stuff is okay. Just don't let it dominate your existence away from the office.

Work Life Versus Home Life: A Constant Struggle

That brings us to the crux of the matter. The fact is, you spend more waking time with the people at your job than you do with your partner. When you think of it that way, it's no wonder so many people feel a sense of rivalry with their partners' jobs:

+ You didn't call. Your dinner's on the table — cold.

+ You said you'd be home at 6:00, and now it's 8:00!

✦ You can't make your daughter's field hockey game?
   She'll be so disappointed. (And so will I.)

✦ What do you mean, you've got work to do? It's 11:30,
   and time for bed.

✦ Your boss wants you to come in on *Saturday*?!

The list goes on and on. And it's true: Work *does* interfere
with home life. The only way to cope successfully with this fact is
for both of you to understand and accept two realities:

✦ Work and time spent on work translate into the income
   you need as a couple.

✦ The partner who is working is also making choices
   whenever possible to be with his or her partner and
   family.

Where problems arise is when one person starts to feel that
his partner's work life is becoming a replacement for home life,
rather than a necessity. To avoid this, both partners have to make
an ongoing effort to be understanding about the competing demands that job and home place on the working partner, and to
work at balancing these demands. Working partners should . . .

✦ **Set a regular, consistent time for leaving work.** As an
   employer, I value the time my employees spend with their
   families. If their family life is positive, then they're going to
   be better, more productive workers. Once in a while, when a
   project needs to be completed, a late evening is highly
   appreciated. But we don't make a habit of it, and we
   certainly don't pull all-nighters anymore. (For one thing,
   everybody is wasted for the next couple of days afterward,
   so productivity actually goes down as a result.) In fact, part

of each person's job is to manage the workflow so that late nights aren't necessary. Any time people *do* have to work late, I expect them to compensate by taking off an extra day or half-day later on.

✦ **Arrive home when expected.** Stopping with coworkers for a drink without calling to tell your partner is just as inconsiderate as staying late at work and not calling.

✦ **Keep your SO informed.** Working partners should call as soon as they know they're going to be late, and should also call just before leaving the office, so their SO knows when they are going to be arriving home.

At the same time, the working partner's SO should . . .

✦ **Make an effort for the partner who is working late.** My wife would always make sure a warm meal was waiting for me when I got home, no matter how late it was.

✦ **Make an effort to be interested.** You don't have to be a rocket scientist to be in a relationship with a rocket scientist. Your partner may not be able to discuss all the details of his or her work, but you can still show empathy by being sincerely interested in how your SO's work went that day, and by being a good listener when he or she responds to your questions or comments.

In some homes both partners work, often long hours. No matter how hard the day has been, when they each get home they need to assess how hard the day has been for the other partner and be ready to provide support if needed.

✦ Consider establishing a time to be in touch with each other in the afternoon. That contact can be the time to decide who stops to pick up dinner or who's going to cook that night.

✦ Schedule a regular date night (see "Date Night, Part 1," page 129). It's all too easy to let work take over to the point where you never see each other.

✦ Make a pact to leave home straightened up so that when you both return home in the evening, you arrive home to an environment that doesn't make you feel like you have three more hours of work to do before you can make dinner.

Staying Home Is Work, Too

———✦———

There was a time when you had to get up on a soapbox and shout for all the world to hear that staying home is work, too. Today, however, it's widely recognized that—between maintaining the household, caring for children, running errands, and perhaps doing some paid freelance work as well—stay-at-home partners work every bit as hard as partners who go off to a job, and they are entitled to the same consideration.

When Joe gets home from his job, he needs to provide just as much support and encouragement for Nancy as she does for him. If he sees that she's had a tough day, he should encourage her to kick back after dinner while he does the dishes. Maybe he can even draw a bath and bring her robe to her, along with a cup of tea or a glass of wine. These gestures work best when they're done with no ulterior motives, but rather out of a desire to be respectful of your partner and the efforts he or she makes for your relationship. Everybody needs a shoulder to lean on once in a while. This is Joe's turn to be that shoulder for Nancy.

By the way, this advice applies not just for stay-at-home moms but for the growing ranks of stay-at-home dads, too.

When Careers Collide

Just because both partners in a relationship work outside of the home doesn't mean that all of the problems of the struggle between work and home disappear. Quite the opposite, in fact. Denny and Pam both have jobs. They also have two children, a dog and a cat, and a house to take care of. If they're not careful, they could be in trouble. Why? Because now both people have jobs and home pulling at them—meaning there's twice as much opportunity for stress to build to the point where it becomes overwhelming. When that happens, look out.

Another potential pitfall is the "my job takes up so much time that I don't have time for my partner" syndrome. I just read the most amazing story in an advice column. It seems that the woman in question's job was several hours away from the home she shared with her husband. She was also a pretty high muckity-muck at work, so she opted to rent an apartment near work to have a place to stay at on weeknights.

You can see where this is heading. It wasn't long before the only time she was coming home was for holidays and long weekends—and when she was home, she was busy trying to take over command and control of household.

You can see where this relationship is heading, too.

I'll be frank with you: I've been there. I let myself get so wrapped up in my work that nothing else mattered. That's a mistake. Work matters, but other things matter more.

You matter.

Your life matters.

Your relationship matters.

Your family matters.

Making the effort to assess your situation and balance your work life with your home life is worth it. I looked at what I was doing and what I was taking on at work, and I discovered that I

> ## ETIQUETTE IMPERATIVE
>
> ### Jobs Don't Love You Back
>
> ———◆———
>
> Never let your job become so important to you that it replaces your partner. Your relationship matters. Fight for it. Communicate, compromise, and repeatedly commit to what you have together. Remember, when you go to bed at night, your job isn't going to hold you in its arms.

was taking responsibility for other people's work. So I met with the people involved and put the responsibility back on them, where it belonged. An example: One evening, I was the one who stayed late to fix a problem with a job and get it out the door on schedule. The next day, I met with the employees working on the project and made it clear that next time I wouldn't fix their problem—they would.

To be fair, periodic spates of long work hours do happen and can be tolerated. If you have your priorities straight and your partner knows this, then your partner will also know that you're making choices with the best interests of your relationship and/or family in mind, and that you will never sacrifice your relationship for work.

The best thing about the work world today is that people no longer have one job for a lifetime. People change jobs and even careers. If the demands of a job grow to the point where they threaten your relationship then a decision has to be made. People used to ask me how I had the gumption to leave a full-time job and start my own ad agency. For me, the harder choice by far would have been to remain in the job I left.

When Your Partner Visits Your Workplace

◆

Be proud. Make a point of introducing your partner to your co-workers, and also to your boss if he or she is available.

If you're busy at the moment your SO shows up, at least acknowledge that you know he or she has arrived, and take a moment to explain that you'll be finished very shortly.

Make your SO feel welcome. How you treat your partner will be noticed not only by him or her but also by your coworkers and will help them frame their opinions of you.

Office Parties and Other Events

Mary and John have been going to John's annual office party for the past nine years. This year, though, Mary is adamant: She's not going. And John can't understand why not.

It's true that last year he'd had to work late on the day of the party, and for good reason: He had screwed up the year before that, and gotten a little hammered at the party. So last year he'd worked hard right through the start of the party, to make sure everybody knew he was determined not to repeat the same mistake again.

And hadn't he already apologized many times over for the year he left her alone the whole evening while he schmoozed with his new supervisor? It really didn't have anything to do with the fact that the new boss was a woman. And anyway, she's no longer working there.

And then there was the year he somehow forgot to tell Mary about the party altogether. Of course, once he realized his oversight he called her right away that afternoon. She had no sitter lined up, and her hair was a mess, but he knew she'd be able to pull things together, and she did. Of course, when they got back home, he had the nerve to complain about how embarrassed he was that she hadn't been more outgoing.

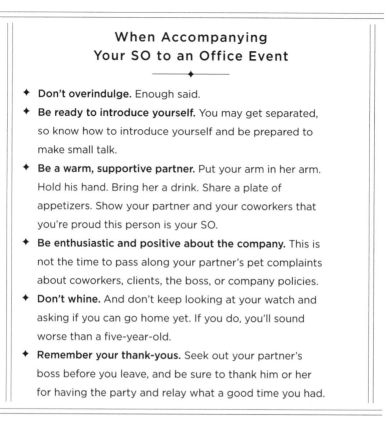

When Accompanying
Your SO to an Office Event

◆

+ **Don't overindulge.** Enough said.

+ **Be ready to introduce yourself.** You may get separated, so know how to introduce yourself and be prepared to make small talk.

+ **Be a warm, supportive partner.** Put your arm in her arm. Hold his hand. Bring her a drink. Share a plate of appetizers. Show your partner and your coworkers that you're proud this person is your SO.

+ **Be enthusiastic and positive about the company.** This is not the time to pass along your partner's pet complaints about coworkers, clients, the boss, or company policies.

+ **Don't whine.** And don't keep looking at your watch and asking if you can go home yet. If you do, you'll sound worse than a five-year-old.

+ **Remember your thank-yous.** Seek out your partner's boss before you leave, and be sure to thank him or her for having the party and relay what a good time you had.

And still he can't understand why this year she refuses to attend.

Business Trips

Every now and then you may be lucky enough (or unlucky enough, depending on your point of view) to accompany your SO on a business trip. Many of the same rules that apply to a company event also apply to business trips. Being engaged and interested in what is going on, making small talk with people you may or may not know, and of course dressing appropriately, all make you look good and also reflect well on your partner.

Remember, pillow talk is powerful. Not only will your partner's coworkers be impressed by your actions, words, and appearance, but their significant others will notice, too. And, because people are naturally gossipy, they will share opinions at night, in the privacy of their rooms.

As a traveling spouse, you may also be expected to attend various events, either with or without your partner. If the trip involves working sessions for employees, companies will often set up events or excursions for their accompanying partners. If so, make the effort to attend. Think of it as an opportunity to get to know the firm's other significant others. Not only will you be helping your partner build better relationships at work but you may even discover you have some future good friends on that bus to the Corn and Alfalfa Museum.

The Home Office

The home office would seem to be the perfect solution to the struggle between work and personal life. And it's true that with a few carefully established guidelines, working from home can relieve a lot of pressure. These guidelines are essential to making the situation succeed, however, and both partners have to respect them.

First, establish a specific place where the working-from-home partner will go to work. I have a little "office" in a hallway area outside the downstairs bathroom. It's about five feet by nine feet—just big enough for a desk, a bookshelf, and an extra table for my printer. From a convenience point of view, it's great: The bathroom is only about two steps from my desk chair.

Second, the most important aspect of this home office space is your attitude and your SO's attitude toward it. When I am in there working, I might as well be on the moon. There isn't even a

========= ETIQUETTE IMPERATIVE =========

Working at Home Is No Vacation

———◆———

Would you go home early from the office to help your partner paint the living room?

Would you come into work late because you were mowing the lawn?

Would you take an extra hour at lunch to watch a soap opera?

Would you leave work early to give your neighbor a helping hand?

No? Then don't do these things just because you're working at home.

---

door to close—but I know that when I'm there, I'm on the job. And so does my wife.

When you are working from home, one thing that helps in staying professional is to be consistent about your schedule. Start work each day at the same time, and follow a regular pattern throughout the day. This type of routine is good for you, and it's also good for your partner—especially if he or she is at home, too. By sticking to a regular schedule, both of you will know when you're at work and when you're available.

I also always recommend to home workers that they dress for the business day. Casual is fine, but this doesn't mean wearing your workout clothes—or worse yet, your bathrobe and slippers. Dressing in a sharp, fresh change of clothes will help put you in the frame of mind that you are "going to work." It will also help your partner (and your neighbors and your pals) realize that you're serious about being at work, and thereby lessen the threat that you'll be interrupted. And if your SO should happen to have a friend drop over unexpectedly, your "office" wardrobe will also spare everyone the discomfort of having you waltz through the house in boxers and a T-shirt or your ratty sweats.

The Partner's Responsibilities

◆

The partner of a work-at-home person has certain responsibilities as well. In particular, the partner shouldn't ask the working SO to interrupt whatever he or she is doing to complete a chore around the house, to answer the home phone, or to do errands. The litmus test is simple: "Would I interrupt my SO for this if he or she were away at work?" If not, then don't do it at home, either.

Example: Sally just called to see if you and your SO can go to dinner next week. Would you drop everything and phone your partner at work to ask if that sounded okay? No, you'd wait till he or she came home that evening to discuss it. So don't disturb your partner in the home office, either.

Vacations and Leisure Time

There's a trend in the modern work world that I find abhorrent—people skipping vacations. And yet, people often feel they need to go that extra mile just to preserve their job.

This is *not* a good thing.

From a business standpoint, it's a fundamental mistake to think that people will continue to perform at a high level when they don't take vacations. An employee's vacation is vital to a business, because it is an opportunity for that worker to recharge and come back to the workplace energized.

When I started my ad agency, I worked like a dog, but each summer I managed to get away to a family vacation home that we share with my brothers and sister. That week was always a godsend.

After about six or seven years of this, for some inexplicable reason we took two weeks. I'll never forget that vacation. It wasn't until somewhere around Thursday of the first week that it dawned on me: I still had more than a week of vacation time in front of me. It was in that second week that I really began to unwind for the first time since I'd started the business.

When I got back to work, I gathered the staff together and told them that I'd come up with some new ideas while on vacation. I could see the concern written all over their faces. *This is fun,* I thought. Then I announced that starting immediately, everyone in the company would be getting three weeks of vacation per year instead of two. The one caveat: Each employee had to take two of the weeks consecutively.

It was easily one of the best business decisions I ever made—and I think my employees would agree.

In addition to vacations being good for you and your business, they're very important for your relationship. In fact, your relationship is much like a busy worker, constantly dealing with the many pressures and demands you and your SO place on it during the course of your daily lives. And like any worker, your relationship needs a vacation—a time for both of you to recharge and to reconnect with each other.

All Work and No Play Makes for a Dull Relationship

Finally, partners need to watch out for each other. If all Joe ever does is come home from work and immediately dive into helping with the chores every night, and then run around on the weekend from sunup to sundown doing still more chores, eventually he's going to burn out. And if all Nancy does is work and take care of the baby, all day every day, with no respite from cooking, cleaning and running errands, eventually *she's* going to burn out. They need to protect each other from this kind of overload by making sure they each set aside some real leisure time—an athletic activity, a couple of hours with the Sunday crossword puzzle, an afternoon visit with a friend, or a morning on the ski slopes.

19

ON THE ROAD

THAT VACATION THE TWO OF YOU HAVE BEEN LOOKING
forward to all year is finally drawing near. But what's it going to
be—eco-trekking in the Amazon or chilling on a Caribbean
beach?

This is exactly the situation TedandDenise find themselves
in: They each have a couple of weeks' vacation time coming to
them, and they've diligently worked the systems at their respective jobs to make sure they'll be able to take their vacations together. Now the time has come when they have to decide what
they're actually going to do with those precious weeks.

If they're like a lot of couples, the opening conversation on
the subject might go something like this:

"Where do you want to go?"

"I don't know. Where do *you* want to go?"

"I don't know. Let's talk about it later."

This sort of discussion is definitely not helpful. If each person doesn't speak up and lay claim to whatever it is that he or she really wants to do, they may end up going nowhere. Also, with this kind of waffling, Ted can't bellyache when he finds himself hiking through a bug-infested tropical rainforest instead of deciding which body part to tan first while a waiter fetches him the first rum punch of the day—and Denise similarly can't complain about how bored she is counting the grains of sand on the beach, when she'd much rather be taking in the flora and fauna of a Brazilian jungle.

The interesting thing about deciding where to vacation is that unlike some of the compromises we've examined in other chapters, this one is really hard to split down the middle. Practically speaking, TedandDenise are not going to trek in South America for a week and then spend another full day (or more) in transit so that they can lounge on an island for the last part of their vacation.

When both partners have differing ideas about the ideal vacation, the best solution is to find a place that offers something for each of you—for example, a tropical beach resort that includes inland excursions, giving Denise the opportunity to go out trekking while Ted hangs out in his beach chair with a good book and that rum punch. Otherwise, one of them will have to compromise, since the only other option is to find a third alternative that doesn't work for either of them.

This compromise can come in several different forms:

✦ I wash your back, and you wash mine. We'll do the eco trek this year, and the Caribbean beach next year.

✦ You go your way, and I'll go mine. With separate vacations, everyone gets what he or she wants — but no one gets each other. That may work for some people, but not for me.

✦ You find a destination with enough alternatives to provide something for everyone. This may sound at first like a tall order, but such destinations are actually limitless, even if you're just traveling domestically: think Cape Cod, the North Carolina shore, the Finger Lakes region in New York, the Great Smoky Mountains of Tennessee, the Florida Keys, Yellowstone National Park, the Grand Canyon, or the tip of Baja California.

Success in finding a compatible vacation really depends on the effort you're both willing to put into finding and researching places until you finally hit on the solution that works for both of you. If Ted takes the lead and really makes this effort, then Denise in turn may need to be a little more accommodating about her desires, and vice versa. Given all the wonderful vacation opportunities out there, with a little work they're virtually certain to find something that meets both of their needs.

Trains, Planes, and Automobiles (Being Considerate Travelers)

Once the vacation plans are set, the departure day can't come soon enough. You and your SO's challenge now is to be sure that when it arrives, you're ready for it.

Packing

My wife and I travel to Italy almost every year. The night before we leave is always a rush of packing. And that's where the first potential for conflict arises.

The Overnight Bag

Whenever you travel, you and your SO should always carry overnight bags containing your jewelry and other valuables as well as a few essential items, so that you won't be left high and dry when your checked bags don't arrive at your destination. These essentials include

+ A change of underwear for each of you
+ A fresh shirt and a lightweight sweater for each of you. (The sweater is particularly handy in midflight, when the cabin gets a little cool.)
+ Toilet kits
+ Traveler's checks and any cash you take
+ Prescription medicines
+ Spare glasses and contact lenses and solution
+ A copy of your itinerary and a photocopy of your passports
+ Reservation confirmations for hotels, tickets, and automobiles

We've learned through hard experience that the odds of our bags making it to the Fiumicino Airport outside Rome at the same time we do are slim to none, especially if there are plane changes involved. Last year, we thought we finally had the system beaten. We took the same airline all the way, making just one change in Philadelphia followed by a direct flight to Rome. With a full five hours to transfer our two bags from one plane to the next, we figured, they couldn't possibly screw up.

They did.

Now, lost bags are a problem everywhere. We're so jaded about misplaced suitcases that we make it a point to bring our must-have items in carry-on bags (see "The Overnight Bag," above), and we always plan on spending at least one night in the city near the airport so that the airline can get our bags to us easily. This time, when the bags didn't show, we dutifully went to the counter where you report lost baggage. In Rome, this counter is especially big, so it must happen there a lot.

Eventually, we got to the front of the line, reported the missing bags, showed our claim numbers, filled out the appropriate forms, and finally left the airport. Amazingly, our bags showed up at our hotel the next morning.

This story leads back to the first point of conflict: packing for the trip. "Let's really try to pack light this year," I'll say to my wife. "How about fitting everything into two carry-on bags each, so we don't have to worry about checking anything? See, mine fits."

I don't know if I say this seriously or just to see my wife's reaction. Either way, we always end up checking our suitcases.

Whatever strategy you employ, the best approach to packing always comes down to two words: *pack light.* Even if you're not trying to squeeze your traveling clothes into a carry-on bag, don't take more than you need. Right now, that extra sweater, trousers, dress, or pair of shoes doesn't seem to add much weight. But when faced with those 139 steps up to the guesthouse on the cliff overlooking the sea, my wife may rue the day she didn't pack a smaller bag.

My goal is simple: By the end of the trip, I don't want to have anything left in my bag that I haven't used. As a result, I pack a little less each time I travel—because I'm always finding an unused something at the bottom of the bag when I get back home.

Don't Miss the Boat

Whether you're going by plane, train, car, or boat, traveling to a destination is always a little stressful. For one thing, there are invariably departure times that have to be met—or else. Wouldn't it be a bummer to arrive at the dock, only to stand there watching as your cruise ship headed out into the Hudson River?

Don't Play the Blame Game

◆

Ted and Denise are *both* responsible for making sure their passports are current, as well as for knowing their departure times, keeping track of any and all tickets, and counting their bags at each waypoint to be certain they have the same number they started with. If they miss the plane or leave a suitcase behind, they're both at fault.

Don't start pointing fingers, because it's not worth it. Instead, if a glitch arises, solve the problem and move on. You'll arrive less stressed, and you'll have a better vacation.

My wife and I actually saw this happen while on a cruise to Bermuda a couple of years ago. As our ocean liner pulled out of Hamilton Harbor, we watched from the transom as a speedboat raced to catch our ship. Several people had somehow managed to miss the sailing time and were lucky to make it back to the cruise ship. I've always wondered how much they paid that speedboat pilot. Meanwhile, talk about an argument-starter: "I *told* you we had to be back at the dock by three o'clock!" "Well, if you hadn't gotten lost on the moped, this never would have happened—but no, *you* can't be bothered to look at a map!" And so on.

Memo to Selves: Have Fun With Each Other

Once you arrive at your destination, the idea is to have fun with each other, to feed the flame between you and kindle it into a roaring fire for a few days.

So forget the world you left behind. Everything will be okay— even the kids, and even the grandparents watching the kids. Put your cell phones away. Unplug that computer you thought you'd use to stay in touch online. Soak in a tub. Order room service.

Enjoy the amenities of the resort, or whatever place you've traveled to.

Most important, take time to slow down and focus on each other. Seize this opportunity to reconnect with the person you want to be with for the rest of your life.

Even on our trips to Italy, my wife and I gear way down. We're not there to see how much sightseeing we can cram into one day. We're there to enjoy the culture, soak up the atmosphere, eat the most wonderful food in the world, and simply enjoy ourselves. One year, we rented an apartment in Rome with a rooftop terrace (which, by the way, was significantly less expensive than a week in a hotel). Both of us could have spent hours just sitting there, reading a book, enjoying the view over the rooftops of the Pantheon and the dome of St. Peter's Basilica in the distance, reveling in being together with nothing more pressing to think about than where to eat dinner that night. And we did.

The Ugly American

Much has been written about Americans' crude manners in dealing with people on vacation, especially when traveling abroad. Those same crude manners are equally abhorrent at home, of course. In truth, however, most "ugly American" behavior occurs not because people are trying to be rude but because they haven't looked in a mirror lately.

This problem is compounded for a couple, since one person's rudeness affects the two of you. If Ted is short and sarcastic with a waitress, the tension and the resulting quality of her service will affect both Ted and Denise—and the resulting bad mood will settle on everybody at the table. The same goes for how you treat the flight attendant, the conductor on the train, and the person behind the counter at the car rental agency.

Try this mental exercise for a minute: Put yourself in their place. Imagine facing dozens (if not hundreds) of pushy, obnoxious customers each day, and then ask yourself what your mood would be like. One of my favorite times to employ this exercise is when a plane flight has been cancelled. When I get to the ticket counter, I smile, and start by saying, "Boy, this is some pain in the neck for you. That last guy was unbelievable."

Instead of bracing to deal with yet another furious passenger, the harried airline representative can now relax, because I've made it clear that we're on the same side. Then I continue, "It looks like all of the remaining direct flights are full. What do you think is the best thing for us to do now?"

We and two other couples spent half an hour working in this way with a customer service representative one stormy February evening at Dulles International Airport in Washington, D.C. We'd missed a connecting flight to Orlando, and the best alternative she could find was to fly us all the way out to Los Angeles, then route us to Orlando two days later. Eventually, we all settled on hopping the shuttle to Boston and then flying back home to Burlington that evening. The amazing thing was, not a single cross word or drop of anger was expressed during the entire half hour. Just seven people searching for a solution. Turns out some of the other people on our flight were stuck in Dulles for two days.

Traveling With Other People

There are times when traveling alone as a couple, just the two of you, is really the right thing—like the trip my wife and I made to Italy for our twenty-fifth wedding anniversary. At other times, traveling with family and friends can be a wonderful way to share a vacation experience. The trick is to strike a balance between being with your friends while still keeping your focus on each other.

Car Etiquette

———◆———

+ **No arguing in the automobile, especially if you're sharing it with another couple.** Remember, the other couple is stuck with you. There's no escaping now, but they'll think long and hard before ever traveling with you again.

+ **Take it easy on the speed.** I've seen too many lead-footed female drivers to hang this one on Ted's shoulders alone. Either way, whoever is at the wheel should make an extra effort to lighten up on the gas pedal and drive smoothly for the comfort of everyone on board.

+ **No smoking.** Enough said.

+ **No drinking.** More than enough said.

+ **Sweep it out.** Before your guest couple gets in the car, clean out the junk-food wrappers and old papers, and wipe down the seats. If necessary, do a quick vacuuming to get rid of Fido's hairs, so that Jane can finish the ride without getting black hair all over her white skirt.

+ **Check the gas and oil.** One of the fastest ways to put a damper on the journey would be to run out of gas on a country back road. We were the passengers of another couple one night when we literally coasted up to the gas pump just as the engine quit because the car had run out of gasoline. That was a close one.

+ **Know exactly how to get where you're going.** Before you leave, go online and get maps and directions for your destination. Not only is it easy to do, but it's a wonderful time-saver and a great way to avoid arguments.

When we went to Italy with Mac and Virginia, we knew Mac was really into bicycling. We also knew that if we fit in some days of cycling through Tuscany, it would make his trip. So we located a bike rental shop in Florence online and reserved some good road bikes. Once we got to Florence, we navigated our way through the ridiculously narrow back streets to the shop, picked up the bikes, and headed out into the countryside for three days.

Now, I must say that if anyone had told me I was going to be renting bikes in Italy and actually enjoying cycling in the hills of Tuscany, I would have told them they were nuts. Yet it turned out to be one the best experiences any of us ever had. We had a ball seeking out gentle hills that wouldn't kill us, and seeing the countryside from the seat of a bicycle provided a whole new way for my wife and me to enjoy Italy. I have Mac and Virginia to thank for that.

Traveling with friends and family can work and be great fun at the same time, especially if you and your SO both take heed of the following:

- ✦ Be prepared to put up with other people's foibles.

- ✦ Be prepared to compromise more than you might want to.

- ✦ Be ready to try activities that might not be your thing. Even if they're a little strange and new to you, do them anyway, with a smile.

- ✦ Talk to everyone.

- ✦ Remember always that you are TedandDenise—and that everything Denise does and everything Ted does reflects on each other, and on the two of you.

20

Special Occasions

"YOUR SISTER-IN-LAW'S BIRTHDAY PARTY? YOU'VE GOT TO be kidding!"

"Your parents want to have dinner with us—again?"

"Your nephew's bar mitzvah? What time do we have to be there?"

"Your aunt and uncle's fiftieth wedding anniversary? Will I have to get dressed up?"

"An engagement party for your brother—the brother neither of us can stand?"

"A funeral for your third cousin once removed? Do we really have to go?"

In any relationship, special occasions crop up in all shapes and sizes. Some will involve friends or business associates, but many, if not most, will revolve around either your family or the

family of your SO. Whatever the reason for the occasion, however, your SO is going to need your support, not your whining.
That means . . .

✦   No arguing about whether you're going to attend

✦   Being ready to leave on time

✦   Being dressed appropriately

✦   Conversing with relatives and friends when you're
     there, and fully participating in the event

These guidelines are particularly important when the occasion
involves your partner's relatives—a favorite source of grumbling
for many people. Remember that, like it or not, they're your SO's
family and they come as part of the package. Remember, too, that
your SO has to make the same compromises for your family.

When you commit to your SO, you also commit to appearing
at these mandatory events. You can either clear these hurdles
standing at your SO's side and emerge looking like a hero or
heroine—or you can go into a pout and set the stage for acri-
mony later on. Whichever way you choose to go, your attitude
not only will affect your SO but also will leave his or her relatives with a lasting impression of you individually, and of the two
of you as a couple.

Bottom line: Given that you *are* going to attend the event,
you might as well make up your mind right now to dress your
best, go willingly, and enjoy yourself to the fullest. If you do, the
day or evening will be a lot more pleasant for you and everyone
else as a result.

Be There and Care

◆

Funerals require very special care on a partner's part. When your SO has suffered the loss of a close family member or friend, the most important thing you can do is to be there for your partner whole-heartedly, with all the cooperation and strength you can muster.

Getting Hitched

It goes on and on: From the first glimmering thoughts that the two of you might want to tie the knot to when you return from your honeymoon to face the task of finally writing all those thank-you notes you swore you'd do before leaving but never got to, getting married is a never-ending set of situations that all need to be adroitly navigated. I sometimes wonder if the process isn't designed to be a sort of trial by fire—throw everything you can at the couple now, for if they can survive the stress of the wedding, there's a reasonable chance their marriage will survive, too.

It all starts, of course, with the proposal. . . .

Where Should You Propose?

There are few events in life as important as the moment of the marriage proposal. This is one time when you want to get the setting exactly right.

Michael and Nicole had their first dinner date at Shelburne Farms, on the banks of Lake Champlain. There they strolled among the flower gardens and watched the sun set over the lake. It was a wonderful evening, one that both of them sensed was the start of something very special. Months later, when it had come clear that their relationship was destined to become per-manent, Michael confided to me that when he proposed to

Nicole he was going to do it in those gardens, because of the memory of that first date.

He did—and she accepted.

They might not have realized it at the time, but what will undoubtedly happen is that they will go back to Shelburne Farms again and again, sometimes to mark a special anniversary, sometimes "just because." And every time they return there, the memories of that special evening when they decided to get married will come flooding back, creating yet another special moment (see "Your Special Place," page 127).

So think carefully about where and how you'll propose to your SO. (Yes, women do ask for the man's hand in marriage, too, and this is absolutely okay.) Make the moment romantic, and make it special—because you'll both remember it for a long, long time.

The Engagement

The engagement ring has been offered and accepted. Now, DanandRose are *engaged*. Their couple status has just risen a few notches, because they've made a commitment to each other to spend the rest of their lives together. This change in status also means that everything Dan or Rose does must now be done with the thought of how it will affect the other person. They may not have figured it out quite yet, but they'll soon learn that all of their choices from here on in affect not only Dan and Rose but also DanandRose.

Meanwhile, the first of several special occasions is already zooming down the road toward them: the engagement party. Both his parents and her parents may want to give one, depending on the circumstances. The best way to approach engagement parties is to step back and let the parents take the lead. This is *their* party to announce the impending wedding of their son or daughter.

It's Still a Good Idea

---◆---

For a younger couple just starting out, once you've proposed and your partner has accepted, talking to both sets of parents about your plans makes a lot of sense. The old tradition of the man going to his intended's father to ask for her hand in marriage is waning. But the new tradition of talking together as a couple to the parents of the bride-to-be is very much alive. It shows respect for both the mother and father and starts the couple along the path of doing things as a couple, rather than as individuals. Finally, it's equally appropriate to have the same talk with the prospective bridegroom's parents.

DanandRose's main responsibility is to be a couple in love and to share their excitement with the people at the party and make their parents proud. So they're careful to do all the things any honored guest should do at a party:

◆ They greet the guests.

◆ They speak with everyone.

◆ They ooh and ahh and say "thank you" if someone does give them a gift. (Although *Emily Post's Wedding Etiquette* book will tell you that gifts aren't expected at an engagement party, Aunt Tilda still may want to bring something to give her favorite niece.)

◆ They hold hands and smooch every now and then, so everyone can see how in love they are.

◆ As the guests depart, they thank everyone for coming.

◆ Finally, they thank their parents for a wonderful evening.

The next day, the betrothed couple shares the chore of writing thank-you notes to the guests—at least those guests who brought gifts. If they want, they can also write notes to the guests who didn't bring gifts, simply to thank them for coming to the party.

Another special thank-you note to their parents would be pretty awesome, too.

An alternative: If they live far away from family, DanandRose can also accept the offer of friends to have an engagement party in their honor. In special circumstances, it's perfectly appropriate for a friend or a relative living nearby to host the party.

Bachelor/Bachelorette Party

The traditional bachelor/bachelorette party is fast going out of style. Instead of viewing this evening as the chance to have one last fling before pledging your life to your intended, modern couples see it as an opportunity to get together with friends and take a much-needed break from the crush of planning their wedding. Viewed in this light, a bachelor/bachelorette party makes terrific sense. My nephew, Casey, arranged and hosted a great bachelor party for his brother, consisting of a two-day canoeing trip down the Connecticut River. They had a blast, and came back relaxed and ready to tackle the wedding.

Michael and Nicole went for another option that couples are choosing more and more often: They combined their parties and went to Montreal for a weekend of eating, partying, and unwinding from the stress of the previous few weeks' activities. That way they were able to enjoy the break and each other at the same time. Personally, I think this tradition beats the "one last fling" approach hands down. The key is to make the party a meaningful addition to your memories. Turn it into a booze fest, on the other hand, and the only memories you'll end up with is who got sick, and how often.

Do It for the Right Reasons

Relegate the traditional "last fling" bachelor (or bachelorette) party to the dustbin of history, where it belongs. This party should be a time to celebrate your upcoming marriage with close friends, and to assure them that your friendships will continue, even though the groom's (or bride's) status is changing.

The Big Day

With all the hoopla swirling around a wedding, it's all too easy to forget that the day is really all about you and your partner—the newly joined DanandRose. Stories are legion of meddling mothers, bickering divorced parents, cantankerous siblings, and pushy aunts, uncles, cousins, and friends who do outrageous things such as bringing uninvited guests to the event.

Whenever I talk with brides and grooms, often with their families right there by their sides, I look them in the eye and tell them to remember two things:

- ✦ This is your day. Be prepared to make decisions and stand by them—even in the face of flak from close family.

- ✦ This day is about you, so keep your focus on each other, and on the vows and commitment you're making together.

You will remember this day for the rest of your life, and so will your partner. When conflicts arise or tempers start fraying, Dan and Rose should step away, take a deep breath, turn to each other and remind themselves of this day's true purpose. Remember, too, that your wedding day is about relationships, not things. First and foremost, it is about your new relationship. You've moved into a new realm, where commitment takes on much greater meaning. You aren't just "going out" anymore; you're committed to building a life together, and sharing that life as one—a couple.

- ✦ **Stay with each other.** Experience the day as a couple, not as two individuals.

- ✦ **Talk with each other.** It's so tempting to talk with everyone else at the event that you can forget to talk with your new spouse. After all, talking is a major part of sharing.

- ✦ **Laugh with each other.** It's important to have fun at your wedding and to let yourselves go. Laughter is the best remedy for any stress that may have been building in the runup to the big day.

- ✦ **Enjoy your guests.** There's always going to be someone present who says or does something unusual (such as giving you the strangest gift in the world or imbibing too much at the rehearsal dinner and acting foolishly). Forget the hassles, and focus on all the good things this person has brought to your life. After all, that's why he or she is attending in the first place.

- ✦ **Thank your in-laws.** Even if difficulties should arise between you and them, remember that they raised the person you've fallen in love with.

- ✦ **Thank your parents.**

Your wedding day is also about your other relationships, including . . .

- ✦ **Your relationship with your parents.** One way or another, they will always be a part of your lives. So help your new relationship—between your parents and the two of you as a couple—start on a positive note.

✦ **Your relationships with brothers, sisters, and other family.**
A sibling or other close relative may feel some jealousy
toward this intruder who is taking away their companion
and confidant. Your goal as a couple is to help that sibling
realize that he or she still is going to retain that relationship,
and that the new partner is an added bonus, not a usurper.

✦ **Your relationships with friends.** Make a point of talking to
your friends during the wedding reception. Let them see how
happy you are together. This is a new dynamic, for them and
for you. Ideally, each friendship will develop into a new
relationship with the two of you. So help your friends see that
even though this is your wedding day, they still matter to you.

The Honeymoon

After all those hectic days leading up to the wedding, when you
and your SO had virtually no time for each other, the honeymoon is a time to finally be alone as a couple, just the two of you.
It's a time to reconnect, and to start making the adjustment to
your new status as husband and wife.

For couples that have been alone together before, either on
trips or because they've already been living together, the honeymoon is a wonderful chance to get away from it all and enjoy
each other. For couples who haven't lived together or traveled together, on the other hand, the honeymoon is that great moment
when you'll experience the first blush of sharing the same room
and waking up together.

No matter how much you try to visualize your honeymoon
ahead of time, the reality will be full of surprises. It will also be a
test of the ties that truly bind you.

For Dan and Rose, who didn't live together before they got
married, those first few days were a revelation. They loved being

alone, knowing they could be together and be intimate, and they had fun sharing meals, going on excursions and lying on the beach. Yet even on their honeymoon, a few clouds inevitably darkened the horizon—issues such as . . .

+ His snoring—or *her* snoring. How will they each get a full night's sleep? (See "Snoring," page 72.)

+ Does Rose leave the bathroom door open, or close it behind her?

+ Do they share the bathroom while he's shaving and showering and she's putting on makeup—or do they use it one at a time?

+ How does Rose handle it when Dan starts throwing up one night after eating something that didn't agree with him? Should she get up and try to comfort him, or pretend she slept right through the whole episode?

+ Should Dan pick up the clothes that Rose has left scattered over the floor and furniture and say nothing, or will this become the subject of their first argument?

Dan and Rose need to be aware that issues like these are inevitably going to arise, and they need to be prepared and committed to resolving such issues together and moving beyond them. The above are just a few of the many little things that will challenge them and their relationship, both right away and in the future, as they continue to grow as a couple. There will be big challenges waiting further down the road as well. But for now, learning to keep their perspective in the days leading up to their wedding, on the big day itself, and on their honeymoon will give DanandRose what they need most: a strong start on the path to being a successful couple.

Afterword

LIKE A BRIEF TROPICAL STORM, THE CLOUDS OVER DANANDROSE move on, leaving clear skies behind. They learn to have patience with each other, to talk things through and give a little on each side. In this way, they start to build their relationship together as DanandRose.

The amazing thing is, this building never stops. It hasn't stopped for my wife and me, even after thirty-two years of marriage, and I can see it still going on with my parents, who have been married for over sixty years. Becoming a couple is a process that continues over time, for as long as you're together.

We are all individuals; each one of us unique. In making the commitment to be a couple and share our lives with one another, however, we also make a commitment to temper our individuality so that we can enjoy being a couple.

Etiquette helps us become that couple by encouraging us to treat each other with consideration, respect, and honesty—the three essential principles that form the basis for a successful relationship. If you look at any one of the many problems discussed in this book, you'll see that invariably it can be traced to the fact that one of these three principles has been neglected or forgotten.

The interesting thing is that when problems *do* crop up, the way out of the conflict is also based on these very same principles. Etiquette will serve as your unfailing guide in forging and constantly renewing that commitment the two of you made when you first decided to become a couple, and in communicating and compromising effectively as you strive to resolve your issues—and in doing all of this with sincerity.

Making your relationship work as a couple requires an ongoing effort, true, but the trade-off is well worth it—for what the two of you are building together adds up to something much, much greater than the sum of the two of you apart.

INDEX

EMILY POST 1872 TO 1960

EMILY POST BEGAN HER CAREER AS A WRITER AT THE AGE OF thirty-one. Her romantic stories of European and American society were serialized in *Vanity Fair, Collier's, McCall's,* and other popular magazines. Many were also successfully published in book form.

Upon its publication in 1922, her book *Etiquette* topped the nonfiction bestseller list, and the phrase "according to Emily Post" soon entered our language as the last word on the subject of social conduct. Mrs. Post, who as a girl had been told that well-bred women should not work, was suddenly a pioneering American career woman. Her numerous books, a syndicated newspaper column, and a regular network radio program made Emily Post a figure of national stature and importance throughout the rest of her life.

> "Good manners reflect something from inside—an innate sense of consideration for others and respect for self."
>
> —Emily Post